D1496470

Universitext

Editors

F.W. Gehring
P.R. Halmos

Universitext

Editors: F.W. Gehring, P.R. Halmos

Booss/Bleecker: Topology and Analysis
Charlap: Bieberbach Groups and Flat Manifolds
Chern: Complex Manifolds Without Potential Theory
Chorin/Marsden: A Mathematical Introduction to Fluid Mechanics
Cohn: A Classical Invitation to Algebraic Numbers and Class Fields
Curtis: Matrix Groups, 2nd ed.
van Dalen: Logic and Structure
Devlin: Fundamentals of Contemporary Set Theory
Edwards: A Formal Background to Mathematics I a/b
Edwards: A Formal Background to Higher Mathematics II a/b
Endler: Valuation Theory
Frauenthal: Mathematical Modeling in Epidemiology
Gardiner: A First Course in Group Theory
Godbillon: Dynamical Systems on Surfaces
Greub: Multilinear Algebra
Hermes: Introduction to Mathematical Logic
Hurwitz/Kritikos: Lectures on Number Theory
Kelly/Matthews: The Non-Euclidean, The Hyperbolic Plane
Kostrikin: Introduction to Algebra
Luecking/Rubel: Complex Analysis: A Functional Analysis Approach
Lu: Singularity Theory and an Introduction to Catastrophe Theory
Marcus: Number Fields
McCarthy: Introduction to Arithmetical Functions
Meyer: Essential Mathematics for Applied Fields
Moise: Introductory Problem Course in Analysis and Topology
Øksendal: Stochastic Differential Equations
Porter/Woods: Extensions of Hausdorff Spaces
Rees: Notes on Geometry
Reisel: Elementary Theory of Metric Spaces
Rey: Introduction to Robust and Quasi-Robust Statistical Methods
Rickart: Natural Function Algebras
Schreiber: Differential Forms
Smith: Power Series from a Computational Point of View
Smoryński: Self-Reference and Modal Logic
Stanisić: The Mathematical Theory of Turbulence
Stroock: An Introduction to the Theory of Large Deviations
Sunder: An Invitation to von Neumann Algebras
Tolle: Optimization Methods

K.T. Smith

Power Series from a Computational Point of View

Springer-Verlag
New York Berlin Heidelberg
London Paris Tokyo

Kennan T. Smith
Mathematics Department
Oregon State University
Corvallis, Oregon 97331, USA

AMS Classification: 26-01

With 2 Illustrations

Library of Congress Cataloging in Publication Data
Smith, Kennan T., 1926–
  Power series from a computational point of view.
  (Universitext)
  Includes index.
  1. Analytic functions. 2. Power series. I. Title.
QA331.S618 1987   515'.2432   87-4854

© 1987 by Springer-Verlag New York Inc.
All rights reserved. This work may not be translated or copied in whole or in part without the written permission of the publisher (Springer-Verlag, 175 Fifth Avenue, New York, New York 10010, USA), except for brief excerpts in connection with reviews or scholarly analysis. Use in connection with any form of information storage and retrieval, electronic adaptation, computer software, or by similar or dissimilar methodology now known or hereafter developed is forbidden.
The use of general descriptive names, trade names, trademarks, etc, in this publication, even if the former are not especially identified, is not to be taken as a sign that such names, as understood by the Trade Marks and Merchandise Marks Acts, may accordingly be used freely by anyone.

Printed and bound by R.R. Donnelley & Sons, Harrisonburg, Virginia
Printed in the United States of America.

9 8 7 6 5 4 3 2 1

ISBN 0-387-96516-5 Springer-Verlag New York Berlin Heidelberg
ISBN 3-540-96516-5 Springer-Verlag Berlin Heidelberg New York

PREFACE

At the end of the typical one quarter course on power series the students lack the means to decide whether $1/(1+x^2)$ has an expansion around any point $\neq 0$, or the tangent has an expansion anywhere — and the means to evaluate and predict errors.

In using power series for computation the main problems are: 1) To predict a priori the number N of terms needed to do the computation with a specified accuracy; and 2) To find the coefficients $a_0, \ldots, a_N$. These are the problems addressed in the book.

Typical computations envisioned are: calculate with error $\leq 10^{-6}$ the integrals

$$\int_0^{\pi/2} (\pi/2-x)\tan x \, dx \,, \quad \int_0^1 x(1+e^x)^{1/2} \, dx,$$

or the solution to the differential equation

$$y''+(\sin x)y'+x^2 y = 0, \quad y(0) = 0, \quad y'(0) = 1,$$

on the interval $0 \leq x \leq 1$.

This computational point of view may seem narrow, but, in fact, such computations require the understanding and use of many of the important theorems of elementary analytic function theory: Cauchy's Integral Theorem, Cauchy's Inequalities, Unique Continuation, Analytic Continuation and the Monodromy Theorem, etc. The computations provide an effective motivation for learning the theorems and a sound basis for understanding them. To other scientists the rationale for the computational point of view might be the need for efficient accurate calculation; to mathematicians it is the motivation for learning theorems and the practice with inequalities, ϵ's, δ's, and N's.

Throughout the book $\epsilon = 10^{-6}$. Experience shows that 10^{-6} (or any other specific small number) is more acceptable and challenging to students than a vague and mysterious ϵ, while, of course, there is no difference in the mathematical analysis. 10^{-6} is chosen so that those who want to can perform realistic computations on a 16 bit microcomputer. The computer code is usually a mathematical proof in a disguise that is appealing to students, and it is strongly recommended as a required part of the problem solutions, simply as a learning device.

Since the book contains complete proofs of the theorems cited above, it is clear that the whole cannot be covered in one quarter. A one quarter course, especially one for engineers, physicists, etc., might cover Chapters 1 and 2 with intensive discussion of the meaning and application of the theorems, but without proofs. (This has been done several times with gratifying success.) A two or three quarter course might cover the whole with proofs and other topics. (The simple proof of the general homotopy version of Cauchy's Theorem was devised in such a course about twenty-five years ago.)

TABLE OF CONTENTS

CHAPTER 1. TAYLOR POLYNOMIALS

1. TAYLOR POLYNOMIALS — 1
2. EXPONENTIALS, SINES, AND COSINES — 4
3. THE GEOMETRIC SUM — 6
4. COMBINATIONS OF TAYLOR POLYNOMIALS — 14
5. COMPLEX TAYLOR POLYNOMIALS — 19
   PROBLEMS — 23

CHAPTER 2. SEQUENCES AND SERIES

1. SEQUENCES OF REAL NUMBERS — 30
2. SEQUENCES OF COMPLEX NUMBERS AND VECTORS — 33
3. SERIES OF REAL AND COMPLEX NUMBERS — 35
4. PICARD'S THEOREM ON DIFFERENTIAL EQUATIONS — 42
5. POWER SERIES — 48
6. ANALYTIC FUNCTIONS — 56
7. PREVIEW — 59
   PROBLEMS — 63

CHAPTER 3. POWER SERIES AND COMPLEX DIFFERENTIABILITY

1. PATHS IN THE COMPLEX PLANE C — 68
2. PATH INTEGRALS — 70
3. CAUCHY'S INTEGRAL THEOREM — 72
4. CAUCHY'S INTEGRAL FORMULA AND INEQUALITIES — 76
   PROBLEMS — 84

CHAPTER 4. LOCAL ANALYTIC FUNCTIONS

1. LOGARITHMS — 87
2. LOCAL SOLUTIONS TO ANALYTIC EQUATIONS — 91
3. ANALYTIC LINEAR DIFFERENTIAL EQUATIONS — 99
   PROBLEMS — 107

CHAPTER 5. ANALYTIC CONTINUATION

1. ANALYTIC CONTINUATION ALONG PATHS 110
2. THE MONODROMY THEOREM 116
3. CAUCHY'S INTEGRAL FORMULA AND THEOREM 122
    PROBLEMS 124
    INDEX 129

CHAPTER 1. TAYLOR POLYNOMIALS

1. TAYLOR'S FORMULA.

THEOREM 1.1 (Mean value theorem) If g and h are continuous on the closed interval I and differentiable on the open interval and a and x are points in I, then there is a point c between a and x such that

$$h'(c)(g(x)-g(a)) = g'(c)(h(x)-h(a)), \quad \text{or}$$

$$\frac{g(x)-g(a)}{h(x)-h(a)} = \frac{g'(c)}{h'(c)} \quad \text{if the denominators are} \neq 0. \quad (1.2)$$

Proof. If

$$F(t) = h(t)(g(x)-g(a))-g(t)(h(x)-h(a))$$

the proof amounts to showing that $F'(c) = 0$ for some c. Substitution shows that $F(x) = F(a)$, therefore that if F is not constant (in which case F' is identically 0), then either the maximum or the minimum of F on $[a,x]$ occurs at an interior point c of $[a,x]$ with $F'(c) = 0$.

Remark 1.3 If $h(x) = x-a$, then (1.2) becomes

$$g(x) = g(a) + g'(c)(x-a) \quad (1.4)$$

which is the usual mean value theorem.

DEFINITION 1.4 If $f, f', \ldots, f^m$ exist at a, the Taylor polynomial of degree m of f at a is the polynomial

$$T_a^m f(x) = \sum_{n=0}^{m} a_n (x-a)^n, \quad a_n = f^n(a)/n! \qquad (1.5)$$

THEOREM 1.6 $T_a^m f$ is the only polynomial P of degree $\leq m$ satisfying

$$P^k(a) = f^k(a), \quad k = 0, \ldots, m \qquad (1.7)$$

It also satisfies $\quad (T_a^m f)' = T_a^{m-1} f' \qquad (1.8)$

Proof. (1.8) is proved by inspecting both sides. (1.7) results from:

If $P(x) = \sum_{n=0}^{m} b_n x$, then $P(x) = \sum_{n=0}^{m} a_n (x-a)^n$,

with $\quad a_n = P^n(a)/n!. \qquad (1.9)$

To see that P has the form on the right, write $x = (x-a)+a$ and expand the powers by the binomial theorem. To verify the formula for the coefficients differentiate P, in its form on the right, n times, put $x = a$, and note that the only non-zero term in the sum is $n! a_n$

DEFINITION 1.10 The difference

$$R_a^m f(x) = f(x) - T_a^m f(x) \qquad (1.11)$$

is called the remainder after degree m. It is the error incurred in using $T_a^m f$ as an approximation to f.

Taylor's formula provides one means of evaluating

the remainder, or error.

THEOREM 1.12 (Taylor's formula) Let $f, f', \ldots, f^{m+1}$ exist on an open interval I. If a and x are two points in I, there is a point c between them such that

$$f(x) = T_a^m f(x) + \frac{f^{m+1}(c)}{(m+1)!} (x-a)^{m+1}, \quad \text{or} \quad (1.13)$$

$$R_a^m f(x) = \frac{f^{m+1}(c)}{(m+1)!} (x-a)^{m+1} \quad (1.14)$$

Proof. For $m = 0$, (1.13) becomes the usual mean value theorem (1.4), and the proof continues by induction, using (1.2) with $g = R_a^m f$ and $h = (x-a)^{m+1}$. Since both g and h vanish at a, there is a point d between a and x such that

$$\frac{R_a^m f(x)}{(x-a)^{m+1}} = \frac{(R_a^m f)'(d)}{(m+1)(d-a)^m} = \frac{R_a^{m-1} f'(d))}{(m+1)(d-a)^m}$$

the second equality coming from (1.8). Now, induction provides a point c between d and a such that

$$R_a^{m-1} f'(d) = \frac{(f')^m (c)}{m!} (d-a)^m = \frac{f^{m+1}(c)}{m!} (d-a)^m .$$

Formula (1.14) results from substituting this in the last formula above.

DEFINITION 1.15 If $R_a^m f(x) \longrightarrow 0$, i.e. if $T_a^m f(x) \longrightarrow f(x)$, (which sometimes can be shown through (1.14)) that fact is expressed by the statement

$$f(x) = \sum_{n=0}^{\infty} a_n(x-a) \quad , \quad a_n = f^n(a)/n!. \qquad (1.16)$$

2. EXPONENTIALS, SINES, AND COSINES

If $E(x) = e^x$, then $E^n(x) = e^x$ for every n, so

$$e^x = \sum_{n=0}^{m} \frac{x^n}{n!} + \frac{e^c x^{m+1}}{(m+1)!}, \quad c \text{ between } 0 \text{ and } x, \qquad (2.1)$$

the polynomial being $T_0^m E$.

If $S(x) = \sin(x)$, then

$$S^{2n}(x) = (-1)^n \sin(x) \text{ and } S^{2n+1}(x) = (-1)^n \cos(x), \quad \text{so}$$

$$\sin(x) = \sum_{n=0}^{m} \frac{(-1)^n x^{2n+1}}{(2n+1)!} + \frac{(-1)^{m+1} \cos(c) x^{2m+3}}{(2m+3)!} \qquad (2.2)$$

In this case the polynomial is both $T_0^{2m+1} S$ and $T_0^{2m+2} S$.

Similarly,

$$\cos(x) = \sum_{n=0}^{m} \frac{(-1)^n x^{2n}}{(2n)!} + \frac{(-1)^{m+1} \cos(c) \, x^{2m+2}}{(2m+2)!} \qquad (2.3)$$

A rough estimate of the factorials is

$$n! > (n/e)^n \qquad (2.4)$$

SECTION 2. EXPONENTIALS, SINES, AND COSINES

For $n = 1$ this says that $1 < 1/e$, and the proof goes by induction. If (2.4) holds, $(n+1)! > (n+1)(n/e)^n$, and the inductive step is ok if $(n+1)(n/e) > ((n+1)/e)^{n+1}$, which is equivalent to $(1+1/n) < e$. For the above functions (2.4) gives

$$\left| R_0^m E(x) \right| \leq M \left(\frac{e|x|}{m+1} \right)^{m+1}, \quad M < e, \qquad (2.5)$$

$$\left| R_0^{2m+1} S(x) \right| \leq \left(\frac{e|x|}{2m+3} \right)^{2m+3}, \qquad (2.6)$$

$$\left| R_0^{2m} C(x) \right| \leq \left(\frac{e|x|}{2m+2} \right)^{2m+2}, \qquad (2.7)$$

According to Definition 1.15, this means that

$$e^x = \sum_{n=0}^{\infty} x^n/n!,$$

$$\sin(x) = \sum_{n=0}^{\infty} \frac{(-1)^n x^{2n+1}}{(2n+1)!} \qquad (2.8)$$

$$\cos(x) = \sum_{n=0}^{\infty} \frac{(-1)^n x^{2n}}{(2n)!}$$

Problem. $T_0^{2m+1} S$ is to be used as an approximation to the sine on the interval $|x| \leq \pi/2$. How large should m be to guarantee an error $\leq 10^{-6}$

6   CHAPTER 1.   TAYLOR POLYNOMIALS

Answer. According to (2.6),

$$\left|S(x)-T_0^{2m+1}S(x)\right| \leq \left(\frac{e\pi}{2(2m+3)}\right)^{2m+3} \qquad (2.9)$$

It is easily seen that the right side is $\leq 10^{-6}$ if m = 5, indeed that

$$\left|S(x)-T_0^{11}S(x)\right| \leq 5.71 \times 10^{-8}, \text{ for } |x| \leq \pi/2. \qquad (2.10)$$

To check out m = 4 it is necessary to use (2.2), which gives

$$\left|S(\pi/2)-T_0^9 S(\pi/2)\right| = \cos(c)(\pi/2)^{11}/11!$$

Since c is unknown, the worst case, $\cos(c) = 1$, must be assumed, in which case the error is 3.5×10^{-6}.

Evaluations on a 16 bit microcomputer give

$$\left|S(\pi/2)-T_0^9 S(\pi/2)\right| = 3.5 \times 10^{-6}$$

$$\left|S(\pi/2)-T_0^{11} S(\pi/2)\right| = 10^{-7}$$

so (2.9) and (2.10) are pretty good.

3. THE GEOMETRIC SUM

If $G(x) = 1/(1-x)$, then $G^n(x) = n!(1-x)^{n+1}$, so

SECTION 3. THE GEOMETRIC SUM

$$\frac{1}{1-x} = \sum_{n=0}^{m} x^n + \frac{x^{m+1}}{(1-c)^{m+2}} \qquad (3.1)$$

$$= T_0^m G(x) + R_0^m G(x).$$

It is not true here that for every x, $R_0^m G(x) \longrightarrow 0$ as $m \longrightarrow \infty$. Replacing m by m+1 in (3.1) and subtracting the two gives

$$0 = T_0^{m+1} G(x) - T_0^m G(x) + R_0^{m+1} G(x) - R_0^m G(x)$$

$$= x^{m+1} + R_0^{m+1} G(x) - R_0^m G(x),$$

so if $R_0^m G(x) \longrightarrow 0$, then $x^{m+1} \longrightarrow 0$, and $|x|$ must be < 1. It is true that if $|x| < 1$, then $R_0^m G(x) \longrightarrow 0$, but this cannot be seen from (3.1). If x > 0, then all that is known about c is that 0 < c < x < 1. This means that 1-c > 1-x, therefore that $|R_0^m G(x)| < x^{m+1}/(1-x)^{m+2}$, which goes to 0 only if x < 1/2. From (3.1) it can be deduced that $R_0^m G(x) \longrightarrow 0$ if -1 < x < 1/2, but no more can be deduced.

While Taylor's formula is fundamental in theoretical work, it has two disadvantages from the point of view of computation: except in special cases the necessary derivatives are awkward to calculate, and even when they can be found, the error estimate may not be adequate because of the unknown point c. In this particular case a better formula than (3.1) is obtained by noticing that $T_o^m G(x) - x T_o^m G(x) = 1 - x^{m+1}$, so that

$$\frac{1}{1-x} = \sum_{n=0}^{m} x^n + \frac{x^{m+1}}{1-x} \qquad (3.2)$$

This shows that $R_0^m G(x) = x^{m+1}/(1-x)$, from which it is clear that $R_0^m G(x) \longrightarrow 0$ if and only if $|x| < 1$. According to Definition 1.5, this means that

$$\frac{1}{1-x} = \sum_{n=0}^{\infty} x^n, \quad \text{for} \quad |x| < 1 \qquad (3.3)$$

Formula (3.2) can also be integrated and differentiated, while (3.1) cannot. Differentiation of (3.2) (with m replaced by m+1) gives

$$\frac{1}{(1-x)^2} = \sum_{n=0}^{m} (n+1) x^n + \frac{x^{m+1}(m+2-(m-1)x)}{(1-x)^2} \qquad (3.4)$$

According to (1.8), the polynomial is the Taylor polynomial of degree m. By l'Hospital's rule, if $|x| < 1$, then, for any k, $m^k x^m \longrightarrow 0$ as $m \longrightarrow \infty$, so if $|x| < 1$, the remainder, $R_0^m(x) \longrightarrow 0$ as $m \longrightarrow \infty$. Consequently,

$$\frac{1}{(1-x)^2} = \sum_{n=0}^{\infty} (n+1) x^n, \quad \text{for } |x| < 1. \qquad (3.5)$$

Integration of (3.2) from 0 to x gives

$$-\log(1-x) = \sum_{n=1}^{m+1} x^n/n + \int_0^x \frac{t^{m+1}}{1-t} dt. \qquad (3.6)$$

According to (1.8), the polynomial and $T_0^{m+1} L$ ($L(x) = -\log(1-x)$) have the same derivative, so differ by a constant. They must be equal, as both are 0 at 0. By the argument just after (3.1), $R_0^m L(x)$ cannot $\longrightarrow 0$ unless $|x| \leq 1$.

For $x > 0$,

$$\left| R_0^{m+1} L(x) \right| \leq \frac{1}{1-x} \int_0^x t^{m+1} dt = \frac{x^{m+2}}{(m+2)(1-x)}. \qquad (3.7)$$

For $x < 0$,

$$\left| R_0^{m+1} L(x) \right| \leq \int_0^x t^{m+1} dt = \frac{|x|^{m+2}}{m+2}, \qquad (3.8)$$

The two show that $R_0^m L(x) \longrightarrow 0$ for $-1 \leq x < 1$, hence that

$$-\log(1-x) = \sum_{n=0}^{\infty} x^n/n, \quad \text{for } -1 \leq x < 1. \qquad (3.9)$$

Replacement of x by $-x^2$ in (3.2) gives

$$\frac{1}{1+x^2} = \sum_{n=0}^{m} (-1)^n x^{2n} + \frac{(-1)^{m+1} x^{2m+2}}{1+x^2} \qquad (3.10)$$

In this case it is not obvious that the polynomial is Taylor's polynomial of degrees 2m and 2m+1. That it is, follows from Theorem 4.4 in the next section. It is obvious that the remainder $\longrightarrow 0$ for $|x| < 1$, hence

$$\frac{1}{1+x^2} = \sum_{n=0}^{\infty} (-1)^n x^{2n}, \quad \text{for } |x| < 1. \qquad (3.11)$$

Integration of (3.10) from 0 to x gives

$\arctan(x) =$

$$\sum_{n=0}^{m} \frac{(-1)^n x^{2n+1}}{2n+1} + (-1)^{m+1} \int_0^x \frac{t^{2m+2}}{1+t^2} dt, \quad \text{with}$$

$$\int_0^x \frac{t^{2m+2}}{1+t^2} dt \leq \int_0^x t^{2m+2} dt \leq \frac{|x|^{2m+3}}{2m+3} \qquad (3.12)$$

Again, it is not obvious that the polynomial is Taylor's polynomial of degrees 2m+1 and 2m+2, as will follow from the next section, but it is obvious that the remainder $\longrightarrow 0$ for $|x| \leq 1$. Therefore,

$$\arctan(x) = \sum_{n=0}^{\infty} \frac{(-1)^n x^{2n+1}}{2n+1}, \quad \text{for } |x| \leq 1. \qquad (3.13)$$

To extend (3.4), fix k and replace m by m+k-1 in

(3.2) to obtain

$$\frac{1}{1-x} = \sum_{n=0}^{m+k-1} x^n + \frac{x^{m+k}}{1-x} \quad .$$

Differentiation k-1 times yields

$$\frac{1}{(1-x)^k} = \sum_{n=0}^{m} \binom{n+k-1}{n} x^n + \frac{x^{m+1}}{1-x} \sum_{n=0}^{k-1} \binom{m+k}{m+n+1} \left(\frac{x}{1-x}\right)^n \quad (3.14)$$

where the polynomial is T_0^m and

$$\binom{p}{q} = \frac{p!}{q!(p-q)!} \quad \text{is the binomial coefficient.} \quad (3.15)$$

It is easily checked that

$$\binom{m+k}{m+n+1} \leq \binom{m+k}{m+1}\binom{k-1}{n} \quad ,$$

therefore that for $|x| < 1$ the sum on the right of (3.14) is less than

$$\binom{m+k}{m+1} \sum_{n=0}^{k-1} \binom{k-1}{n} \left(\frac{|x|}{(1-|x|)}\right)^n = \binom{m+k}{m+1} \frac{1}{(1-|x|)^{k-1}} \quad .$$

Thus, the remainder in (3.14) satisfies

$$|R_0^m(x)| \leq \binom{m+k}{m+1} \frac{|x|^{m+1}}{(1-|x|)^k} \leq \frac{(m+k)^{k-1}|x|^{m+1}}{(1-|x|)^k}$$

If $|x| < 1$, $R_0^m(x) \to 0$ as $m \to \infty$, so that

$$\frac{1}{(1-x)^k} = \sum_{n=0}^{\infty} \binom{n+k-1}{n} x^n, \quad \text{for } |x| < 1. \qquad (3.17)$$

To treat the function $1/(b-z)^k$ and the Taylor center a, write

$$\frac{1}{b-z} = \frac{1}{b-a} \frac{1}{1-x} \quad \text{with } x = \frac{z-a}{b-a}$$

and substitute in (3.14) to get

$$\frac{1}{(b-z)^k} = \frac{1}{(b-a)^k} \sum_{n=0}^{m} \binom{n+k-1}{n} \left(\frac{z-a}{b-a}\right)^n + R_a^m(z) \qquad (3.18)$$

with $R_a^m(z) = R_0^m(x)/(b-a)^k$. In this case the polynomial is T_a^m, and $R_a^m(z)$ goes to 0 as $m \to \infty$ if $|z-a| < |b-a|$. Therefore, if $|z-a| < |b-a|$,

SECTION 3. THE GEOMETRIC SUM

$$\frac{1}{(b-z)^k} = \frac{1}{(b-a)^k} \sum_{n=0}^{\infty} \binom{n+k-1}{n} \left(\frac{z-a}{b-a}\right)^n . \quad (3.19)$$

If both sides of (3.14) are multiplied by $(1-x)^k$, the result is a polynomial identity that holds for all real x - therefore, for all complex x. Consequently, (3.14) holds for all complex $x \neq 1$, and (3.18) holds for all complex $z \neq b$.

THEOREM 3.20 Let f be a rational function, i.e. $f = P/Q$ with P and Q polynomials, and let $Q(a) \neq 0$. Then $R_a^m f(z) \longrightarrow 0$ as $m \longrightarrow \infty$ if $|z-a| < r$, where r is the distance from a to the nearest complex zero of Q. If Q can be factored, $T_a^m f$ and $R_a^m f$ can be obtained from (3.18) and (3.14).

Proof. By the partial fraction theorem, f is a polynomial plus terms of the form $c/(b-z)^k$, where b is a zero of Q. The results above can be applied to each such term.

Of course, it would not be expedient to use the Taylor polynomial simply to compute the values of a rational function. But, it can be efficient to compute a more complicated function, of which a rational function is a part, by forming the Taylor polynomial of the whole. For example, see Problem 2, section 2, Chapter 1.

4. COMBINATIONS OF TAYLOR POLYNOMIALS

DEFINITION 4.1 If f and g are defined near a, $f = g \mod(x-a)^{m+1}$ means that

$$\lim_{x \to a} \frac{f(x)-g(x)}{(x-a)^m} = 0$$

If f and g are polynomials in x-a, this means that the the terms of degree $\leq m$ are the same in f and g, i.e. that f-g is divisible by $(x-a)^{m+1}$.

THEOREM 4.3 Let $f, f', \ldots, f^m$ exist at a, and let Q be a polynomial. Then $f = Q \mod(x-a)^{m+1}$ if and only if $Q = T_a^m f \mod(x-a)^{m+1}$.

Proof. It is shown first that $f = T_a^m f \mod (x-a)^{m+1}$. For m = 1 this is the definition of the derivative. If m > 1, apply l'Hospital's rule m-1 times to $g = f - T_a^m f$ and $h(x) = (x-a)^m$, and use (1.8) to get to m = 1. (Note that the existence of f^m at a implies the existence of f^{m-1} on an interval with center a, therefore the continuity of earlier derivatives on such an interval.) It is clear that if $f = g \mod (x-a)^{m+1}$ and $g = h \mod(x-a)^{m+1}$, then $f = h \mod(x-a)^{m+1}$.

THEOREM 4.4 a) $T_a^m(f+g) = T_a^m f + T_a^m g$

b) $T_a^m(fg) = T_a^m f T_a^m g \quad \mod(x-a)^{m+1}$.

c) $T_a^m(f \circ g) = T_b^m f \circ T_a^m g \quad \mod(x-a)^{m+1}$, with $b = g(a)$.

SECTION 4. COMBINATIONS OF TAYLOR POLYNOMIALS 15

Proof. a) is obvious from the fact that $(f+g)^n = f^n + g^n$. Now

$$fg - T_a^m f T_a^m g = T_a^m f R_a^m g + T_a^m g R_a^m f + R_a^m f R_a^m g.$$

By Theorem 4.3, each term on the right $= 0 \mod (x-a)^{m+1}$. To prove c), let $\epsilon > 0$ be given and set $M = |g'(a)| + 1$. First choose η so that for $|y-b| < \eta$,

$$|f(y) - T_b^m f(y)| \leq \frac{\epsilon}{2M^m} |y-b|^m . \qquad (4.5)$$

Let M' be the maximum of $|(T_b^m f)'|$ on $|y-b| < \eta$. By the mean value theorem

$$|T_b^m f(z) - T_b^m f(w)| \leq M' |z-w| \text{ if } |z-b| \,\&\, |w-b| < \eta. \quad (4.6)$$

Now choose δ so that $M'\delta < \eta$ and so that if $|x-a| < \delta$, then

$$|g(x) - b| \leq M |x-a| ,$$

$$|T_a^m g(x) - b| \leq M |x-a| , \qquad (4.7)$$

$$|g(x) - T_b^m g(x)| \leq \frac{\epsilon}{2M'} |x-a|^m \qquad (4.8)$$

If $|x-a| < \delta$, then by (4.5) and (4.7)

$$|f(g(x)) - T_a^m f(g(x))| \leq \frac{\epsilon}{2M^m} |g(x) - b|^m \leq \frac{\epsilon}{2} |x-a|^m$$

and by (4.6), (4.7) and (4.8)

$$\left|T_a^m f(g(x))-T_a^m f(T^m g(x))\right| \leq M'\left|g(x)-T_a^m g(x)\right| \leq \frac{\epsilon}{2}|x-a|^m$$

The two together show that

$$\left|f(g(x))-T_a^m f(T_a^m g(x))\right| \leq \epsilon|x-a|^m \quad \text{if } |x-a| < \delta.$$

Since ϵ is arbitrary, this proves c).

Part b) of the theorem says that the Taylor polynomial of degree m of a product can be obtained by multiplying the Taylor polynomials of degree m of the factors, and discarding the terms of degree $> m$. Here there is a simple formula for the coefficients. Let

$$f = \sum_{n=0}^{m} a_n(x-a)^n \mod(x-a)^{m+1} \tag{4.9}$$

$$g = \sum_{n=0}^{m} b_n(x-a)^n \mod(x-a)^{m+1} \tag{4.10}$$

$$h = \sum_{n=0}^{m} c_n(x-a)^n \mod(x-a)^{m+1} \tag{4.11}$$

If $h = fg$, then

$$c_n = \sum_{k=0}^{n} a_k b_{n-k}, \tag{4.12}$$

as follows from multiplying the polynomials and collecting coefficients. Formula (4.12) also provides

SECTION 4. COMBINATIONS OF TAYLOR POLYNOMIALS

a way to find the coefficients of $f = h/g$ when the coefficients of g and h are known and $b_0 = g(a) \neq 0$. If $n = 0$, (4.12) reads $c_0 = a_0 b_0$, so that $a_0 = c_0/b_0$. If $a_0, \ldots, a_{n-1}$ have been found, (4.12) gives

$$c_n = \sum_{k=0}^{n-1} a_k b_{n-k} + a_n b_0 \qquad (4.13)$$

which determines a_n.

In general, (4.13) does not provide a convenient explicit formula for all a_n, but only a means for finding a particular a_n when the previous ones are known.

Part c) of Theorem 4.4 says that to obtain the Taylor polynomial of degree m of the composite function $h(x) = f(g(x))$ at the point a, let $b = g(a)$, find the Taylor polynomials $T_a^m g(x)$ and $T_b^m f(y)$, substitute $y = T_a^m g(x)$ in the latter, and discard the terms of degree $> m$. In this case there are explicit formulas for the coefficients, but they are complicated, and again it is useful to be able to determine in advance how big to take m in order to guarantee a prescribed allowable error. Effective error evaluations are given in the following sections.

Incidentally, formula (4.12) is Leibnitz's formula for the derivative of a product:

$$(fg)^n = \sum_{k=0}^{n} \binom{n}{k} f^k g^{n-k} \tag{4.14}$$

since $a_k = f^k(a)/k!$ etc.

THEOREM 4.15 If w satisfies the "algebroid" equation

$$w^n + \sum_{k=0}^{n-1} a_k(x) w^k = 0 \tag{4.16}$$

then $T_a^m w$ satisfies the corresponding equation

$$(T_a^m w)^n + \sum_{k=0}^{n-1} (T_a^m a_k)(T_a^m w)^k = 0 \quad \mod(x-a)^{m+1}. \tag{4.17}$$

Proof. This is clear from Theorem 4.4.

THEOREM 4.18 If y satisfies the differential equation

$$y^d = \sum_{k=0}^{d-1} a_k(x) y^k + f \tag{4.19}$$

then $T_a^{m+d} y$ satisfies the corresponding equation

$$(T_a^{m+d} y)^d = \sum_{k=0}^{d-1} (T_a^m a_k)(T_a^{m+d} y)^k + f \quad \mod(x-a)^{m+1}. \tag{4.20}$$

5. COMPLEX TAYLOR POLYNOMIALS.

Theorem 3.20 suggests that the possibility of approximating a function by its Taylor polynomials depends on the behavior of the function in the complex domain, rather than just its behavior in the real domain. For example, there is nothing suspicious about the real function $1/(1+x^2)$, but the singularities of the complex function $1/(1+z^2)$ at $z = \pm i$ limit the Taylor approximations centered at 0 to $|x| < 1$. Henceforth, all functions are complex valued functions of the complex variable z unless stated otherwise.

DEFINITION 5.1 Let g be defined on some disk with center a except perhaps at a. The statement

$$\lim_{z \to a} g(z) = w$$

means that for each positive number ϵ there is a positive number δ such that $|g(z)-w| < \epsilon$ whenever $|z-a| < \delta$ and $z \neq a$. The usual limit rules are the same in the real and complex cases, and they are proved in exactly the same way.

THEOREM 5.2 If $\lim_{z \to a} g_1(z) = w_1$, and $\lim_{z \to a} g_2(z) = w_2$ then

$$\lim_{z \to a} (g_1+g_2)(z) = w_1+w_2, \qquad \lim_{z \to a} (g_1 g_2)(z) = w_1 w_2,$$

$$\lim_{z \to a} (g_1/g_2)(z) = w_1/w_2 \quad \text{provided } w_2 \neq 0.$$

DEFINITION 5.3 Let f be defined on some disk with center a. f is complex differentiable at a if the limit

$$f'(a) = \lim_{z \to a} \frac{f(z)-f(a)}{z-a} \quad \text{exists.}$$

If the limit does exist, the complex number f'(a), is the complex derivative of f at a.

The usual differentiation rules are the same in the real and complex cases, and they are proved in the same way.

THEOREM 5.4 If f and g are complex differentiable at a, then f+g and fg are complex differentiable at a, and so is f/g if g(a) ≠ 0. If g is complex differentiable at a, and f is complex differentiable at b = g(a), the composite f∘g(z) = f(g(z)) is complex differentiable at a. As usual,

$$(f+g)'(a) = f'(a)+g'(a), \quad (fg)'(a) = f'(a)g(a)+f(a)g'(a)$$

$$(f/g)'(a) = (g(a)f'(a)-g'(a)f(a))/g(a)^2,$$

$$(f \circ g)'(a) = f'(b)g'(a).$$

The theorem shows that polynomials are

SECTION 5. COMPLEX TAYLOR POLYNOMIALS

differentiable at every point, and rational functions (quotients of polynomials) are differentiable at all points where the denominator is $\neq 0$. It will be shown presently that the common functions such as e^z, $\sin(z)$, $\cos(z)$, etc. can be defined for complex z, and that they are differentiable at every point. The theorem then shows that the various combinations of these are differentiable. In short, it provides quite a variety of complex differentiable functions.

Complex differentiability is nevertheless much more restrictive than real differentiability. The simple function $f(z) = \bar{z}$ is not complex differentiable at any point. Indeed, $(f(a+h)-f(a))/h = \bar{h}/h$, which can be any complex number of absolute value 1, for arbitrarily small $|h|$.

When the required complex derivatives exist, the Taylor polynomials are defined as before.

DEFINITION 5.5 If $f, f', \ldots, f^m$ exist at a, the Taylor polynomials are defined as before.

$$T_a^m f(z) = \sum_{n=0}^{m} a_n (z-a)^n \text{ with } a_n = \frac{f^n(a)}{n!} . \quad (5.6)$$

In the form of Theorem 1.12, Taylor's formula is false in the complex domain. (Once the sine and cosine are defined, it will be easy to see that if $f(z) = \cos(z) + i\sin(z)$, then $f'(z) = -\sin(z) + i\cos(z)$ is $\neq 0$ for all z - so there is no point c with $0 = f(2\pi)-f(0) = f'(c)(2\pi-0)$.)

On the other hand, the theorems on combinations in the last section are all true, and the proofs are all correct except for a small gap. The usual proof of l'Hospital's rule, which was used in the proof of Theorem 4.3 is based on the mean value theorem - just seen to be false in the complex case. The gap is closed when the theory is available in Chapter 3.

CHAPTER 1. TAYLOR POLYNOMIALS - PROBLEMS

PROBLEMS

SECTION 1.

1. Find $T_0^5 f$ for the following functions f.

   a) $1/(1-x)$ g) $e^x/(1-x)$
   b) $\log(1-x)$ h) $e^x \log(1-x)$
   c) e^x i) e^{2x}
   d) $\sin(x)$ j) $e^x \sin(x)$
   e) $\cos(x)$ k) $e^x \cos(x)$
   f) $\tan(x)$ l) $e^x \tan(x)$

2. Estimate the error in using $T_0^5 f$ as an approximation to the f in Problem 1, a) - f) on

   a) $0 \leq x \leq \pi/8$ b) $0 \leq x \leq \pi/4$.

3. Find $T_{\pi/8}^5 f$ for the f in Problem 1.

4. Estimate the error in using $T_{\pi/8}^5 f$ as an approximation to the f in Problem 1, a) - f) on $0 \leq x \leq \pi/4$. Compare with the result of 2b).

SECTION 2.

1. On $0 \leq x \leq \pi/2$, $S(x) = \sin(x)$ is to be computed via $T_a^m S(x)$, where $a = 0$, $\pi/6$, $\pi/3$, or $\pi/2$, depending on which is closest to x.

   a) Write $T_a^m S$ for each of the above centers a. How large must m be to ensure error $\leq 10^{-6}$ on $0 \leq x \leq \pi/2$?

b) Write the computer program to do the computation when the input is x (i.e. the computer finds a).

c) Compare the efficiency of using these several centers vs. that of using a = 0 for all x.

d) If necessary, revise the computer program to take advantage of the fact that the error depends on x: for each x have the computer stop at error $\leq 10^{-6}$ for that x. Run the program for x = multiples of .05 listing the stopping point m.

2. The integral $\int_0^1 \frac{\sin(x)}{x} dx$ is to be computed.

a) Find m so that $\int_0^1 \frac{R_0^{2m+1} S(x)}{x} dx \leq 10^{-6}$

b) Use $\int_0^1 \frac{T_0^{2m+1} S(x)}{x} dx$ for the answer. It follows from section 4 that $\frac{T_0^{2m+1} S(x)}{x} = T_0^{2m} \frac{S(x)}{x}$, but this is not needed here.

c) Compare the efficiency of this integration with that of a Riemann sum, trapezoid, or Simpson integration of the same function.

d) Re-do the problem with Taylor center a = $\pi/8$, and compare efficiencies.

3. Draw conclusions about Taylor centers.

SECTION 3.

1. Compute $\log(z)$ for $1 \leq z \leq 10$, or equivalently $\log(1+x)$ for $0 \leq x \leq 9$, with an error $\leq 10^{-6}$, via the steps below.

a) Write the geometric sum formula for $F(x) = 1/(1+x)$ centered at the point a, and integrate from a to x to obtain

$$\log(1+x) - \log(1+a) = \int_a^x T_a^m F(t)\,dt + \int_a^x R_a^m F(t)\,dt.$$

According to the next section, this is Taylor's formula for $\log(1+x)$ centered at a, but this is not needed. The polynomial part will be the approximation, the (absolute value of the) rest, the error. Hint. Write

$$\frac{1}{1+x} = \frac{1}{1+a+x-a} = \frac{1}{1+a} \cdot \frac{1}{1+(x-a)/(1+a)}.$$

Use the geometric sum for $1/(1+z)$ with $z = (x-a)/(1+a)$.

b) Verify that the error goes to 0 if $|x-a| < 1+a$, as is to be expected from Theorem 3.20.

c) Find m so that if $|x-a|/(1+a) \leq 1/10$, then the error is $\leq 10^{-6}$.

d) Find a sequence of intervals I_k with centers a_k and the following properties.

1) The left end point of I_1 is 0.

2) The left endpoint of I_{k+1} is the right end point of I_k, so that each x is in some I_k (and only one if not an end point).

3) If $x \in I_k$, then $|x-a|/(1+a_k) \leq 1/10$, so the error is $\leq 10^{-6}$ with $a = a_k$ and $x \in I_k$.

Hint. Since 0 is the left end point of I_1, $a_1/1+a_1) = 1/10$, it follows that $a_1 = 1/9$. When a_k is found, the right end point b_k of I_k satisfies $(b_k - a_k)/(1+a_k) = 1/10$, so $b_k = (11/10)a_k + 1/10$, and $(a_{k+1} - b_k)/(1+a_{k+a}) = 1/10$ imply that $a_{k+1} = (11/9)a_k + 2/9$.

e) Find the formula for a_k and the k's needed to compute $\log(1+x)$ for $0 \leq x \leq 9$.

f) Write the FORTRAN program to

   1) Compute (successively) a_k and $\log(1+a_k)$ for use in a).

   2) With input x, $0 \leq x \leq 9$, find k so that $x \in I_k$.

   3) Compute $\log(1+x)$ with the polynomial part of the formula in a).

g) Compare the efficiency of this procedure with that of using $T_5^m F$ for all x, $0 \leq x \leq 9$.

(On the author's 16 bit microcomputer this algorithm gives results that agree up to 10^{-6} with the computer's double precision intrinsic function. The single precision intrinsic function is off by at least 3×10^{-6}.)

2. Find $T_0^m f$ for

a) $f(x) = x^2/(x^2-1)$

b) $f(x) = x^2/((x^2-1)(x-3))$

c) $f(x) = (x^2+1)/((x-1)^2(x+2))$

d) In each case, decide for which x, $T_0^m f(x) \to f(x)$.

e) In each case, give an evaluation of the error: If $T_0^m f(x) \to f(x)$ for $|x| \leq r$, find m so that the error is $\leq 10^{-6}$ for $x \leq r/2$.

f) In each case, compute the integral of $f(x)\sin(x)$ from 0 to 1/2 with an error $\leq 10^{-6}$, and compare the efficiency with that of the Riemann sum, trapezoid, and Simpson procedures. Hint: Use $T_0^m f T_0^n S$ as an approximation to fS, determining m and n so the error in the product is admissible.

g) For each f, do a)-e) for $T_4^m f$.

3. Compute $\int_0^1 \arctan(x^2)\,dx$ with error $\leq 10^{-6}$, via the steps below.

a) Write the geometric sum for $1/(1+t)$ centered at 0, put $t = x^2$, and integrate to get a formula for $\arctan(x^2)$. How large must m be to give error $\leq 10^{-6}$ in the integral?

b) Write the geometric sum for $1/(1+t)$ centered at 1/2, put $t = x^2$, and integrate to get a formula for $\arctan(x^2)$. Is it a Taylor's formula? How large must m be in this case?

c) Write Taylor's formula for $1/(1+x^2)$ centered at $1/2$ and integrate to get a formula for $\arctan(x^2)$. How large must m be in this case? Hint:

$$\frac{1}{1+x^2} = \frac{i}{2}\left(\frac{1}{x+i} - \frac{1}{x-i}\right).$$

SECTION 4.

1. Find $T_0^5 f$ for each of the following functions and for the product and quotient of each pair.

$e^{\sin(x)}$, $e^{\cos(x)}$, $\sin(e^x)$, $1/(1+\cos(x))$, $1/(1+\sin^2(x))$, $\cot(x)$

2. Find $T_0^3 w$ for $w(x) = \cos(x) + \sqrt{\cos^2 x - 4\sin(x)}$.

Hint: w satisfies $w^2 - \cos(x)w + \sin(x) = 0$. Use Theorem 4.15.

3. Find $T_0^3 y$, where y satisfies the differential equation

$y'' - \cos(x)y' + \sin(x)y = e^x$ and $y(0) = y'(0) = 0$.

4. How does formula (4.12) imply Leibnitz's formula (4.17)?

5. Re-do problem 1, g)-h), section 1 with present methods and compare the efficacy of these methods with that of the definition of T_0^5.

SECTION 5.

1. Find $T_0^m f$ for $f(x) = 1/1+x^2$)

   a) Try it directly.

   b) Try it by writing $\dfrac{1}{1+x^2} = \dfrac{i}{2}\left(\dfrac{1}{x+i} - \dfrac{1}{x-i}\right)$

2. Re-do problem 2, section 3, replacing x^2-1 by x^2+1. Try it with and without the use of complexes.

CHAPTER 2. SEQUENCES AND SERIES

1. SEQUENCES OF REAL NUMBERS.

While the properties of real power series are based on properties of complex power series, properties of complex numerical series fall back on properties of real numerical series.

DEFINITION 1.1 The sequence $\{x_n\}$ of real numbers converges to the real number x, written $x_n \to x$, or $\lim_{n \to \infty} x_n = x$, or $\lim x_n = x$, if for each positive number ϵ there is an index N such that $|x-x_n| \leq \epsilon$ for all $n \geq N$. The sequence is Cauchy if for each positive number ϵ there is an index N such that $|x_n - x_m| \leq \epsilon$ for all $n, m \geq N$.

The basic elementary theorem is that a sequence converges if and only if it is Cauchy. The point of this theorem is to give a criterion for convergence that does not require a priori knowledge of the limit, which may not be known.

THEOREM 1.2 If $x_n \to x$, then $\{x_n\}$ is Cauchy.
Proof. Given $\epsilon > 0$, choose N so that if $n \geq N$, then $|x-x_n| \leq \epsilon/2$. If n and m are both $\geq N$, then $|x_n - x_m| \leq |x_n - x| + |x - x_m| \leq \epsilon/2 + \epsilon/2 = \epsilon$.

THEOREM 1.3 If $\{x_n\}$ is Cauchy, then $\{x_n\}$ is bounded.

SECTION 1. SEQUENCES OF REAL NUMBERS

Proof. Find N so that $|x_n - x_m| \leq 1$ for $n, m \geq N$, hence so that $|x_n| \leq |x_N| + 1$ for $n \geq N$. Then, for all n, $|x_n|$ is $\leq$ the maximum of $|x_N| + 1$ and the first N terms.

The converse of Theorem 1.2 depends on a fundamental property of the real numbers.

AXIOM 1.4 Every bounded set X of real numbers has a least upper bound $\bar{x} = \sup X$ satisfying the following.
a) For each $x \in X$, $x \leq \bar{x}$.
b) If $b < \bar{x}$, then there exist points $x \in X$ with $x > b$.

Property a) says that $\bar{x}$ is an upper bound for the numbers in X, while b) says that all other upper bounds are larger. Axiom 1.4 is obviously equivalent to

AXIOM 1.5 Every bounded set X of real numbers has a greatest lower bound $\underline{x} = \inf X$ satisfying
a) For each $x \in X$, $x \geq \underline{x}$.
b) If $b > \underline{x}$, then there exist points $x \in X$ with $x < b$.

Property a) says that $\underline{x}$ is a lower bound for the numbers in X, while b) says that all other lower bounds are smaller.

DEFINITION 1.6 Let $\{x_n\}$ be a bounded sequence of real numbers. The numbers $\bar{x} = \limsup x_n$ and $\underline{x} = \liminf x_n$ are defined as follows

$$\bar{x} = \inf \{\bar{x}_m\}, \quad \text{where } \bar{x}_m = \sup \{x_n : n \geq m\} \quad (1.7)$$

$$\underline{x} = \sup \{\underline{x}_m\}, \quad \text{where } \underline{x}_m = \inf \{x_n : n \geq m\} \quad (1.8)$$

The lim sup has the following properties.

a) If $b > \bar{x}$, there is an index N
with $x_n \leq b$ for all $n \geq N$. (1.9)

b) If $b < \bar{x}$, then for any N
there exist $n \geq N$ with $x_n > b$. (1.10)

To prove a), note that for some N it must be true that $\bar{x}_N < b$, hence that $x_n < b$ for all $n \geq N$. To prove b), note that if $x_n \leq b$ for all $n \geq N$, then $\bar{x} \leq x_N \leq b$.

The lim inf has the following properties.

a) If $b < \underline{x}$, there is an index N
with $x_n \geq b$ for all $n \geq N$. (1.11)

b) If $b > \underline{x}$, then for any N
there exist $n \geq N$ with $x_n \leq b$. (1.12)

It is evident that for any bounded sequence $\{x_n\}$

$$\liminf x_n \leq \limsup x_n .$$ (1.13)

THEOREM 1.14 The following are equivalent.
a) $\{x_n\}$ is Cauchy.
b) $\liminf x_n = \limsup x_n$.
c) $\{x_n\}$ converges.

Proof. Theorem 1.2 says that c) implies a), so it is enough to show that a) implies b) and b) implies c). Suppose that a) holds, let $\epsilon > 0$ be given, and

SECTION 2. SEQUENCES OF COMPLEX NUMBERS AND VECTORS   33

choose N so that $|x_n - x_m| < \epsilon$ for $n, m \geq N$. Then $\bar{x} \leq \bar{x}_N \leq x_N + \epsilon$, and $\underline{x} \geq \underline{x}_N \geq x_N - \epsilon$. Therefore, $\bar{x} \leq \underline{x} + 2\epsilon$. Since this holds for all $\epsilon > 0$, $\bar{x} \leq \underline{x}$. Suppose that b) holds and let $\epsilon > 0$ be given. Choose N_1 so that (1.8) holds with $b = \bar{x} + (\epsilon/2)$, i.e. so that $x_n \leq \bar{x} + (\epsilon/2)$ for all $n \geq N_1$, and choose N_2 so that (1.11) holds with $b = \underline{x} - (\epsilon/2)$, i.e. $x_n \geq \underline{x} - (\epsilon/2)$ for all $n \geq N_2$. Then take N to the the larger of N_1 and N_2.

THEOREM 1.15 A bounded nondecreasing sequence of real numbers always converges to its least upper bound. A bounded nonincreasing sequence of real numbers always converges to its greatest lower bound.

Proof. Let $\{x_n\}$ be nondecreasing, and let $\bar{x}$ be the sup. For any $\epsilon > 0$ $\bar{x} - \epsilon$ is no longer an upper bound, so there exists N with $x_N > \bar{x} - \epsilon$. For any $n \geq N$, $x_N \leq x_n \leq \bar{x}$, so $0 \leq \bar{x} - x_n \leq \epsilon$.

2. SEQUENCES OF COMPLEX NUMBERS AND VECTORS.

DEFINITION 2.1 The sequence $\{z_n\}$ of complex numbers converges to the complex number z, written $z_n \rightarrow z$, or $\lim z_n = z$, if for each positive number ϵ there is an index N such that $|z - z_n| \leq \epsilon$ for all $n \geq N$. The sequence is Cauchy if for each $\epsilon > 0$ there is an index N such that $|z_n - z_m| \leq \epsilon$ for all $n, m \geq N$.

THEOREM 2.2 Let $z_n = x_n + iy_n$, $z = x+iy$, with the x's and y's real. Then $\{z_n\}$ is Cauchy if and only if $\{x_n\}$ and $\{y_n\}$ are Cauchy; $z_n \longrightarrow z$ if and only if $x_n \longrightarrow x$ and $y_n \longrightarrow y$.

Proof. Both statements follow from the fact that if $z = x+iy$ and $w = u+iv$ then $\max(|x-u|,|y-v|) \leq |z-w| \leq \sqrt{2}\max(|x-u|,|y-v|)$.

THEOREM 2.3 A sequence of complex numbers converges if and only if it is Cauchy.

Proof. This follows immediately from Theorems 2.2 and 1.14: if $\{z_n\}$ is Cauchy, both $\{x_n\}$ and $\{y_n\}$ are Cauchy. Therefore both $\{x_n\}$ and $\{y_n\}$ converge, and so $\{z_n\}$ converges.

DEFINITION 2.4 If $z = (z_1,\ldots,z_m)$ is a real or complex vector, the absolute value of z is the number

$$|z| = \left(\sum_{i=1}^{m} |z_i|^2\right)^{1/2} \tag{2.5}$$

and the distance between z and w is the number $|z-w|$. A sequence $\{z_n\}$ of vectors converges to the vector z if for each positive number ϵ there is an index N such that $|z-z_n| \leq \epsilon$ for all $n \geq N$. The sequence is Cauchy if for each positive number ϵ there is an index N such that $|z_n - z_m| \leq \epsilon$ for all $n,m \geq N$.

SECTION 3. SERIES OF REAL AND COMPLEX NUMBERS   35

THEOREM 2.6 The sequence $\{z_n\}$ is Cauchy if and only if, for each coordinate i, the sequence $\{(z_n)_i\}$ is Cauchy, and $z_n \longrightarrow z$ if and only if, for each coordinate i, $(z_n)_i \longrightarrow z_i$ - so Cauchy = convergent, as for numbers.

Proof. $\max |z_i| \leq |z| \leq \sqrt{n} \max |z_i|$.

3. SERIES OF REAL AND COMPLEX NUMBERS.

DEFINITION 3.1 If $\{a_k\}$ is a sequence of complex numbers starting with the index k = K, the n-th partial sum of the sequence is the number

$$S_n = \sum_{k=K}^{n} a_k \qquad (3.2)$$

The series $\sum a_k$ is the sequence $\{S_n\}$, and the series converges if this sequence converges, in which case it is said that the infinite sum exists and has the value

$$\sum_{k=K}^{\infty} a_k = \lim_{n \to \infty} S_n .$$

The following two theorems are obvious from the definition.

THEOREM 3.3 Let $\{a_k\}$ start with 0, and let N be arbitrary. If one of the sums

$$S = \sum_{n=0}^{\infty} a_k \quad \text{and} \quad \sum_{n=N+1}^{\infty} a_k$$

exists, so does the other, and $S-S_N$ is the sum on the right.

THEOREM 3.4 If two of the three below exist, so does the third, and the equality holds.

$$\sum_{n=K}^{\infty} (a_k + b_k) = \sum_{n=K}^{\infty} a_k + \sum_{n=K}^{\infty} b_k \qquad (3.5)$$

THEOREM 3.6 If $\sum |a_k|$ converges, so does $\sum a_k$, and if S and T are the sums, then

$$|S-S_N| \leq \sum_{k=N+1}^{\infty} |a_k| = T - T_N . \qquad (3.7)$$

Proof. For $n \geq N$, replace S by S_n in (3.7), and run the sum to n. This version of (3.7) is obviously true, and it implies (3.7) as it stands.

DEFINITION 3.8 If the series $\sum |a_n|$ converges, the series $\sum a_n$ is said to converge absolutely. (Theorem 3.6 says that absolutely convergent series converge. At the level of these lectures there are very few theorems about series that converge, but do not converge absolutely, and such series are not discussed.)

SECTION 3. SERIES OF REAL AND COMPLEX NUMBERS 37

THEOREM 3.9 $\sum |a_n|$ converges if and only if the partial sums are bounded.

Proof. The partial sums are nondecreasing, so by Theorems 1.3 and 1.15 they converge if and only if they are bounded.

THEOREM 3.10 (Integral Test) For $n \geq N$, let $|a_n| = f(n)$, where f is a nonincreasing function on $x \geq N$. Then $\sum |a_n|$ converges if and only if the integral $\int_N^\infty f(x)dx$ exists.

Proof. The integral exists if and only if

$$\lim_{R \to \infty} F(R) \text{ exists, where } F(R) = \int_N^R f(x)dx$$

and the value of the integral is the limit on the left. When $f \geq 0$, F is a nondecreasing function of R, so the limit exists if and only if F is bounded. The theorem follows from the inequalities

$$\sum_{n=N+1}^M |a_n| \leq \int_N^M f(x)dx \leq \sum_{n=N}^M |a_n| \qquad (3.11)$$

which come from the fact that $|a_{n+1}| \leq f(x) \leq |a_n|$ for $n \geq x \geq n+1$. As $M \to \infty$, (3.11) becomes

$$\sum_{n=N+1}^\infty |a_n| \leq \int_N^\infty f(x)dx \leq \sum_{n=N}^\infty |a_n| \qquad (3.12)$$

For series with positive terms, (3.12) can provide a very effective evaluation of the error incurred in using the partial sum S_N as an approximation to the infinite sum S: the integral is an upper bound for $S-S_N$, and a lower bound for $S-S_{N-1}$.

Example. Let $a_n = n^{-p}$, $f(x) = x^{-p}$. It is easily checked that

$$\int_N^\infty f(x)dx = N^{1-p}/(p-1) \qquad \text{if } p > 1 \qquad (3.13)$$

and that the integral does not exist if $p \leq 1$. Therefore, the series converges if and only if $p > 1$, in which case $N^{1-p}/(p-1)$ is a good evaluation of $S-S_N$.

Example. Let $a_n = 1/(n\log^q n)$, $f(x) = 1/(x\log^q x)$. In this case, if $q > 1$,

$$\int_N^\infty \frac{dx}{x\log^q x} = \int_{\log N}^\infty x^{-q}\, dx = \frac{\log^{1-q} N}{q-1} \qquad (3.14)$$

and the integral does not exist if $q \leq 1$. Hence, the series converges if and only if $q > 1$, in which case the right side of (3.14) is a good approximation to $S-S_N$.

Example. Let $a_n = n^3 e^{-n}$, $f(x) = x^3 e^{-x}$. Integration by parts gives

SECTION 3. SERIES OF REAL AND COMPLEX NUMBERS   39

$$\int_N^\infty x^3 e^{-x} dx = e^{-N}(N^3 + 3N^2 + 6N + 6) \qquad (3.15)$$

so the series converges, and the right side is a good estimate of $S - S_N$.

THEOREM 3.16 (Comparison Test) If there are an index N and a number M such that $|a_n| \leq M b_n$ for $n \geq N$, and if $\sum b_n$ converges, then $\sum a_n$ converges absolutely.

Proof. It is plain that if the partial sums of $\sum b_n$ are bounded, so are the partial sums of $\sum |a_n|$.

Example. Let $a_n = P(n)/Q(n)$, where P and Q are polynomials with leading terms pz^r and qz^s, respectively. If $r > s$, $|a_n| \to \infty$, and the series does not converge. If $r = s$, $a_n \to p/q$, and again the series does not converge. If $r < s-1$, and $M > p/q$, there is an index N such that for $n \geq N$, $|a_n| \leq M n^{r-s}$, and the series converges absolutely by comparison with n^{r-s}.

Exercise. Use the above and comparison with $\sum 1/n$ to show that the series does not converge if $r = s-1$. (Don't forget that the coefficients in P and Q are complex.)

Example. Let $a_n = \log^q n / n^p$, $p > 1$ and $q > 0$. Since $\log y^r / y = r \log y / y \to 0$ for all $r > 0$, it follows that for any $s > 0$, there is an index N such that if $n \geq N$, then $\log n \leq n^s$. Fix s small enough so that p-sq > 1. The series converges absolutely by comparison with $1/n^{p-sq}$.

THEOREM 3.17 If $\sum a_n$ converges, then $a_n \to 0$.

Proof. $a_n = S_n - S_{n-1}$. If $\{S_n\}$, converges, then $S_n - S_{n-1} \to 0$ because $\{S_n\}$ is Cauchy.

THEOREM 3.18 (Ratio Test) If $\limsup (|a_{n+1}|/|a_n|) < 1$, then $\sum a_n$ converges absolutely. If $\liminf (|a_{n+1}|/|a_n|) > 1$, then $\sum a_n$ diverges.

Proof. If $r_o < 1$ is the lim sup, let $r_o < r < 1$. There is an index N so that for $n \geq N$, $|a_{n+1}|/|a_n| \leq r$. Successive application of this inequality to $n = N$, $n = N+1, \ldots$, gives $|a_n| \leq |a_N| r^{-N} r^n$, and the series converges by comparison with the geometric series. If $r_o > 1$ is the lim inf, let $r_o > r > 1$. There is an index N so that for $n \geq N$, $|a_{n+1}|/|a_n| \geq r$. In this case, successive application gives $|a_n| \geq |a_N| r^{n-N}$, and $|a_n|$ is unbounded, so cannot $\to 0$.

THEOREM 3.19 (nth Root Test) Let $r_o = \limsup |a_n|^{1/n}$. If $r_o < 1$, the series $\sum a_n$ converges absolutely; if $r_o > 1$ it diverges.

Proof. If $r_o < 1$, let $r_o < r < 1$. There is an index N such that for $n \geq N$, $|a_n|^{1/n} \leq r$, hence $|a_n| \leq r^n$,

SECTION 3. SERIES OF REAL AND COMPLEX NUMBERS 41

so the series converges by comparison with the geometric series. If $r_0 > 1$, let $r_0 > r > 1$. For infinitely many n, $|a_n| > r^n$, and again the terms are unbounded.

THEOREM 3.20 (alternating series) If a_n is real and $a_n \searrow 0$, then $(-1)^n a_n$ converges, and $|S - S_m| \leq a_{m+1}$.

Proof. For $n > m$,

$$(-1)^{m+1}(S_n - S_m) = (a_{m+1} - a_{m+2}) + (a_{m+3} - a_{m+4}) + \ldots \ldots a_n$$

$$= a_{m+1} - (a_{m+2} - a_{m+3}) - (a_{m+4} - a_{m+5}) - \ldots \ldots a_n .$$

Since the expressions in parentheses and any left over terms are ≥ 0, it follows that

$$0 \leq (-1)^{m+1}(S_n - S_m) \leq a_{m+1} .$$

Theorem 3.20 provides good error evaluations for series like those for the sine and cosine, and simple examples of series that converge, but not absolutely (e.g. $\sum (-1)^n / n$).

Example. Find $\int_0^1 \sin(x)/x \, dx$ with error 10^{-10}.

According to formula (2.8) of Chapter 1,

$$\frac{\sin(x)}{x} = \sum_{n=0}^{\infty} \frac{(-1)^n x^{2n}}{(2n+1)!}$$

so, according to Theorem 4.9 (in the next section),

$$\int_0^y \frac{\sin(x)}{x}\, dx = \sum_{n=0}^{\infty} \frac{(-1)^n y^{2n+1}}{(2n+1)(2n+1)!} \qquad (3.21)$$

and, in particular,

$$\int_0^1 \frac{\sin(x)}{x}\, dx = \sum_{n=0}^{\infty} \frac{(-1)^n}{(2n+1)(2n+1)!} \qquad (3.22)$$

By Theorem 3.20, the error $|S-S_m|$ in (3.22) is at most $1/(2m+3)(2m+3)!$, and this is $\leqq 10^{-10}$ if $m = 5$. Therefore,

$$\int_0^1 \frac{\sin(x)}{x}\, dx = \sum_{n=0}^{5} \frac{(-1)^n}{(2n+1)(2n+1)!} \quad \text{with error} \quad 10^{-10}.$$

With a trapezoidal numerical integration the error with n points has the order $1/n^2$, so a trapezoidal integration would require the computation of $\sin(x)/x$ at 10^5 points. With Simpson integration, the error has the order $1/n^4$, and Simpson integration would require the computation of $\sin(x)/x$ at $10^{2.5}$ points.

4. PICARD'S THEOREM ON DIFFERENTIAL EQUATIONS.

The following is the simplest version of the theorem of Picard on the existence and uniqueness of solutions to the (vector) differential equation

$$y'(x) = F(x,y(x)), \qquad y(x_o) = y_o, \qquad x \in (a,b) \qquad (4.1)$$

SECTION 4. PICARD'S THEOREM ON DIFF. EQUATIONS

THEOREM 4.2 (Picard's Theorem) Let F be a continuous vector valued function on $[a,b] \times R^n$ which satisfies

$$|F(t,u)-F(t,v)| \leq M|u-v|$$

for all $t \in [a,b]$ and all $u,v \in R^n$ \hfill (4.3)

For any given $x_0 \in [a,b]$ and any given $y_0 \in R^n$ there is one and only one continuous vector valued function y for which (4.1) holds.

Proof. (4.1) is equivalent to

$$y(x) = y_0 + \int_{x_0}^{x} F(t,y(t))\, dt . \tag{4.4}$$

Define the sequence $\{y_n\}$ inductively by

$$y_n(x) = y_0 + \int_{x_0}^{x} F(t,y_{n-1}(t))\, dt. \tag{4.5}$$

If M_0 is the maximum of $|F(t,y_0)|$ (and $x > x_0$), then

$$|y_1(x) - y_0| =$$

$$\left| \int_{x_0}^{x} F(t,y_0)\, dt \right| \leq \int_{x_0}^{x} |F(t,y_0)|\, dt \leq M_0(x-x_0) .$$

$$|y_2(x)-y_1(x)| \leq \int_{x_o}^{x} |F(t,y_1(t)-F(t,y_o)| \, dt$$

$$\leq M \int_{x_o}^{x} |y_1(t)-y_o| \, dt \leq MM_o \int_{x_o}^{x} (t-x_o)dt = \frac{MM_o(x-x_o)^2}{2}.$$

$$|y_3(x)-y_2(x)| \leq \int_{x_o}^{x} |F(t,y_2(t)-F(t,y_1(t)| \, dt$$

$$\leq M \int_{x_o}^{x} |y_2(t)-y_1(t)| \, dt \leq M^2 M_o \int_{x_o}^{x} (t-x_o)^2/2 \, dt =$$

$$\frac{M^2 M_o (x-x_o)^3}{3!}.$$

It follows that in general

$$|y_n(x)-y_{n-1}(x)| \leq M^{n-1} M_o |x-x_o|^n/n!$$

$$\leq \frac{M^{n-1} M_o (b-a)^n}{n!}. \tag{4.6}$$

which can be verified easily by induction (for $x < x_o$, as well as $x > x_o$).

SECTION 4. PICARD'S THEOREM ON DIFF. EQUATIONS

For any two indices P and N, P > N,

$$|y_P(x)-y_N(x)| \leq \sum_{n=N+1}^{P} |y_n(x)-y_{n-1}(x)|$$

$$\leq (M_o/M) \sum_{n=N+1}^{\infty} \frac{(M(b-a))^n}{n!} .$$

The series on the right is the remainder in the series for $e^{M(b-a)}$, which is known to go to 0 by (2.5) of Chapter 1. This implies that for each x the sequence $\{y_n(x)\}$ is Cauchy, therefore has a limit $y(x)$, and moreover,

$$|y(x)-y_N(x)| \leq (M_o/M) \sum_{n=N+1}^{\infty} \frac{(M(b-a))^n}{n!} . \qquad (4.7)$$

This inequality implies not only that $y_n(x) \to y(x)$ for each individual x, but that $y_n \to y$ uniformly in the sense of the following definition.

DEFINITION 4.8 $y_n \to y$ uniformly on [a,b] if for each $\epsilon > 0$, there is an index N such that $|y(x)-y_n(x)| \leq \epsilon$ for all $n \geq N$ and all $x \in [a,b]$.

A theorem on uniform convergence will be stated, with the aid of this theorem the proof of Picard's theorem will be finished, then the theorem on uniform convergence will be proved.

THEOREM 4.9 (Uniform Convergence) Let $f_n \to f$

uniformly on [a,b] with each f_n continuous on [a,b]. Then f is continuous on [a,b] and

$$\int_a^b f(t)\, dt = \lim_{n \to \infty} \int_a^b f_n(t)\, dt. \qquad (4.10)$$

End of Picard proof. Since

$$|F(t,y(t))-F(t,y_n(t))| \leq M\, |y(t)-y_n(t)|,$$

it follows from (4.7) that $F(t,y_n(t)) \longrightarrow F(t,y(t))$ uniformly on $[x_o, x]$, therefore that $F(t,y(t))$ is continuous and that (4.10) holds for $f_n(t) = F(t,(y_n(t))$ and $f(t) = F(t,y(t))$. The required (4.4) comes from applying this in (4.5).

For the uniqueness, let z also satisfy (4.4). If $u = |y-z|$, then

$$0 \leq u(x) \leq M \int_{x_o}^x u(t)\, dt. \qquad (4.11)$$

If M_x is the maximum of u on $[x_o, x]$, (4.11) implies that

$$M_x \leq M_x M |x - x_o|,$$

therefore that $u = 0$ on $[x_o, x_o+(1/M)]$. Now apply the same argument with x_o replaced by $x_1 = x_o+(1/M)$, then with x_1 replaced by $x_2 = x_1+(1/M)$, etc.

Proof of Theorem 4.9. First the continuity. Let

SECTION 4. PICARD'S THEOREM ON DIFF. EQUATIONS   47

t_o and ϵ be given, and choose N so that $|f(t)-f_n(t)| \leq \epsilon/3$ for all t and all $n \geq N$, then δ so that

$$|f_N(t)-f_N(t_o)| \leq \epsilon/3 \text{ if } |t-t_o| \leq \delta .$$

If $|t-t_o| \leq \delta$, then

$$|f(t)-f(t_o)| \leq$$

$$|f(t)-f_N(t)| + |f_N(t)-f_N(t_o)| + |f_N(t_o)-f(t_o)|,$$

and all three terms on the right are $\leq \epsilon/3$.

To see (4.10), for $\epsilon > 0$, choose N so that $|f(t)-f_n(t)| \leq \epsilon/(b-a)$ for all t and all $n \geq N$. If $n \geq N$, then

$$\left| \int_a^b f(t)\, dt - \int_a^b f_n(t)\, dt \right| \leq$$

$$\int_a^b |f(t)-f_n(t)|\, dt \leq (b-a)\, \epsilon/(b-a).$$

REMARK 4.12 The same proof is valid if the values of the functions y and F are complex vectors instead of real vectors. x, however, is real.

An equation of order n for a real or complex valued function u,

$$u^n(x) = f(x,u(x),u'(x),\ldots,u^{n-1}(x)) , \qquad (4.13)$$

can always be converted to a vector equation of order 1. If u satisfies (4.13), then the vector y defined by $y_i = u^{i-1}$ satisfies

$y'= F(x,y)$ with $F_i(x,y) = y_{i+1}$ for $i < n$,

$F_n(x,y) = f(x,y_1,\ldots,y_n)$.

Conversely, if y satisfies this equation, then $u = y_1$ satisfies (4.13).

The most common equations in which the "Lipschitz" condition (4.3) is satisfied are the linear equations, those of the form

$$y'(x) = A(x)y(x) + f(x) \qquad (4.14)$$

where A is an n x n matrix of continuous functions and f is continuous. In this case (4.3) is satisfied with

$$M^2 \leq \max \sum_{i,j=1}^{n} |a_{ij}(x)|^2 \qquad (4.15)$$

5. POWER SERIES.

A power series is a series of the form

$$f(z) = \sum_{n=0}^{\infty} a_n (z-a)^n \qquad (5.1)$$

where a, z, and the a_n are complex. For each fixed z it is a numerical series like those in Chapter 2. It is understood that f is defined to be the sum of the series at the points z where the series converges, and is undefined elsewhere. Define the number r by

$$1/r = \limsup |a_n|^{1/n} \qquad (5.2)$$

with the convention that $r = 0$ if $|a_n|^{1/n}$ is unbounded.

THEOREM 5.3 The series (5.1) converges absolutely if $|z-a| < r$, and diverges if $|z-a| > r$. r is called the radius of convergence of the series and the disk $|z-a| < r$ is called the disk of convergence.

Proof. If $r = 0$, the series cannot converge for any $z \neq a$. Indeed, if $z \neq a$, fix $b < |z-a|$. Since $|a_n|^{1/n}$ is unbounded, $|a_n| > 1/b^n$ for infinitely many n, and $|a_n||z-a|^n > (|z-a|/b)^n$, for infinitely many n, so $|a_n||z-a|^n$ is unbounded.

If $r > 0$,

$$\limsup (|a_n||z-a|^n)^{1/n} = \limsup |a_n|^{1/n} |z-a|$$

$$= |z-a|/r,$$

and the result follows from the nth root test, Theorem 3.19.

LEMMA 5.4 For any $r_o < r$, there is a number M such that

$$|a_n| \leq M/r_o^n \quad \text{for all n} \tag{5.5}$$

$$n|a_n| \leq M/r_o^n \quad \text{for all n} \tag{5.6}$$

Proof. Since
$\lim \sup (n|a_n|)^{1/n} = \lim n^{1/n} \lim \sup |a_n|^{1/n} = 1/r < 1/r_o$, there is an index N such that $n|a_n| \leq 1/r_o^n$ for $n > N$. Take M to be the max of 1 and the $n|a_n|r_o^n$ with $n \leq N$.

When the M in (5.5) can be estimated, (5.5) gives a uniform error evaluation on smaller disks. Estimates are provided by Cauchy's Inequalities (Chap. 3, Sec. 4).

Fix any $r_1 < r$ and choose r_o between the two. If $|z-a| \leq r_1$, then by the lemma, $|a_n||z-a|^n \leq M(r_1/r_o)^n$, so that

$$|S(z) - S_N(z)| \leq \sum_{n=N+1}^{\infty} |a_n||z-a|^n \leq \frac{Mr_o}{r_1 - r_o} \left(\frac{r_1}{r_o}\right)^{N+1} \tag{5.7}$$

This shows that

$$S_N \to S \text{ uniformly on } |z-a| \leq r_1, \text{ for any } r_1 < r, \tag{5.8}$$

and gives a uniform error estimate on $|z-a| \leq r_1$.

THEOREM 5.9 The function $f(z)$ defined by (5.1) is complex differentiable on $|z-a| < r$, and the derivative is given by

$$f'(z) = \sum_{n=1}^{\infty} n a_n (z-a)^{n-1} \quad \text{on } |z-a| < r. \tag{5.10}$$

Proof. Formula (5.6) shows that the differentiated series (5.10) has the same radius of convergence, so Theorem 5.3 guarantees the absolute convergence of (5.10) on $|z-a| < r$. Temporarily, call the sum g.

It simplifies the notation to assume that $a = 0$, and this assumption is made in the rest of the proof. Fix z with $|z| < r$, and choose r_0 and r_1 with $|z| < r_1 < r_0 < r$. Let $\delta = r_1 - |z|$. For every h with $h < \delta$,

$$\frac{(z+h)^n - z^n}{h} = \sum_{k=0}^{n-1} \binom{n}{k} z^k h^{n-k-1} \leq$$

$$n(|z| + |h|)^{n-1} < n r_1^{n-1} \tag{5.11}$$

the first inequality coming from the fact that

$$\binom{n}{k} = \frac{n}{n-k} \binom{n-1}{k} \leq n \binom{n-1}{k}.$$

According to (5.11) and Lemma 5.4

$$\sum_{n=m+1}^{\infty} \left| a_n \frac{(z+h)^n - z^n}{h} \right| \leq M \sum_{n=m+1}^{\infty} \left(\frac{r_1}{r_0} \right)^n \quad \text{for all}$$

$|h| < \delta$, and

$$\sum_{n=m+1}^{\infty} |na_n z^{n-1}| \leq \frac{M}{r} \sum_{n=m+1}^{\infty} \left(\frac{r_1}{r_o}\right)^n \qquad (5.12)$$

Given $\epsilon > 0$, first fix m so that the right sides of (5.12) are less than ϵ, hence so that

$$\left|\frac{f(z+h)-f(z)}{h} - g(z)\right| \leq$$

$$\left|\sum_{n=0}^{m} a_n \frac{(z+h)^n - z^n}{h} - \sum_{n=1}^{m} na_n z^{n-1}\right| + 2\epsilon$$

for all $|h| < \delta$. By the usual formula for differentiating a finite sum, the large term on the right goes to 0 as $h \to 0$. Therefore, as $h \to 0$, the lim sup of the left side is $\leq 2\epsilon$, for every $\epsilon > 0$, hence is 0.

THEOREM 5.13 The function $f(z)$ defined by (5.1) has complex derivatives of all orders, given by

$$f^k(z) = \sum_{n=k}^{\infty} \frac{n!}{(n-k)!} a_n (z-a)^{n-k}, \text{ for } |z-a| < r, \qquad (5.14)$$

where all of the series have the same radius of convergence r. Therefore, the series is the Taylor series, with the coefficients given by

$$a_k = \frac{f^k(a)}{k!} \qquad (5.15)$$

Proof. Formula (5.14) comes from repeated application of Theorem (5.9) and formula (5.10). Formula (5.15) comes from the fact that if $z = a$ in (5.14), the only nonzero term on the right is the one with $n = k$, i.e. is $k! a_k$.

THEOREM 5.16 Let f be defined by (5.1) for $|z-a| < r$. If $|b-a| < r$,

$$f(z) = \sum_{p=0}^{\infty} b_p (z-b)^p \quad \text{for} \quad |z-b| < r - |b-a| \qquad (5.17)$$

with $b_p = \sum_{n=p}^{\infty} a_n \binom{n}{p} (b-a)^{n-p} =$

$$\sum_{n=0}^{\infty} a_{n+p} \binom{n+p}{n} (b-a)^n . \qquad (5.18)$$

This theorem is an immediate consequence of Theorem 5.9 and Theorem 4.5 of Chapter 3. The present elementary proof is an example of an important method called the Method of Majorants, in which the objective is to produce a simple known function "worse" than the function at hand.

Proof. Given z, fix $r_o < r$ so that $|z-b| < r_o - |b-a|$. According to Lemma 5.4 and formula (3.17) of Chapter 1,

$$|b_p| \leq \frac{M}{r_o^p} \sum_{n=0}^{\infty} \binom{n+p}{n} \left(\frac{|b-a|}{r_o}\right)^n = \frac{Mr_o}{(r_o-|b-a|)^{p+1}} \quad (5.19)$$

This shows that the series for b_p converges and that the series in (5.17) converges if $|z-b| < r_o - |b-a|$. It remains to show that the sum is $f(z)$.

In $T_a^N f$, write $z-a = (z-b)+(b-a)$ and expand $(z-a)^n$ by the binomial theorem to get

$$T_a^N f(z) = \sum_{p=0}^{N} (z-b)^p \left(\sum_{n=0}^{N-p} a_{n+p} \binom{n+p}{n} (b-a)^n \right), \text{ therefore}$$

$$\sum_{p=0}^{N} b_p (z-b)^p - T_a^N f(z) =$$

$$\sum_{p=0}^{N} (z-b)^p \left(\sum_{n=N-p+1}^{\infty} a_{n+p} \binom{n+p}{n} (b-a)^n \right) \quad (5.20)$$

By Lemma 5.4 and formula (3.16) of Chapter 1, the absolute value of the inner sum on the right is bounded by

$$Mr_o \binom{N+1}{p} \left(\frac{|b-a|}{r_o}\right)^{N-p+1} (r_o-|b-a|)^{-p-1} .$$

Replacement of the inner sum by this expression and use of the binomial theorem give

$$\left| \sum_{n=0}^{N} b_p(z-b)^p - T_a^N f(z) \right| \leq \left(\frac{|b-a|}{r_o} + \frac{|z-b|}{r_o-|b-a|} \right)^{N+1} \quad (5.21)$$

Since $|b-a| < r_o$, if $|z-b|$ is sufficiently small, the number in parentheses on the right is < 1, and the left side goes to 0 as $N \longrightarrow \infty$. Since the series in (5.17) converges and since $T_a^N f(z) \longrightarrow f(z)$, this implies that the series converges to $f(z)$ on a sufficiently small disk around b, and in particular that the sum on the left of (5.20) is $T_b^N f$. Thus, (5.20) becomes

$$T_b^N f(z) - T_a^N f(z) = \sum_{p=0}^{N} (z-b)^p \left(\sum_{n=N-p+1} a_{n+p} \binom{n+p}{p} (b-a)^n \right) \quad (5.22)$$

Taking $c = |b-a|$ and $x = c+|z-b|$, apply this to the function

$$g(x) = M \sum_{n=0} \left(\frac{x}{r_o} \right)^n = \frac{Mr_o}{r_o-x}$$

to get

$$T_c^N g(x) - T_0^N g(x) = M \sum_{p=0}^{N} \left(\frac{x-c}{r_o} \right)^p \sum_{n=N-p+1}^{\infty} \binom{n+p}{p} \left(\frac{c}{r_o} \right)^n \quad (5.23)$$

for $|c| < r_o$. From (5.22) and Lemma 5.4, it follows that

$$|T_b^N f(z) - T_a^N f(z)| \leq M |T_c^N g(x) - T_0^N g(x)| \qquad (5.24)$$

and by (3.18) and (3.19) in Chapter 1, the right side $\longrightarrow 0$ for $|x| < r_0$. (g is the "majorant".)

6. ANALYTIC FUNCTIONS.

DEFINITION 6.1 A set G in the plane is open if for each $a \in G$, there is number $r > 0$ such that the disks $|z-a| < r$ is contained in G.

DEFINITION 6.2 An open set G is connected if each two points in G can be joined by a path in G. A path in G from a to b is a continuous function p from an interval $[t_i, t_f]$ into G with $p(t_i) = a$ and $p(t_f) = b$. In this case, a and b are the initial and final points of the path.

(For general sets, the definition of connected is different, and the above is called path connected. For open sets in R^n connected and path connected are equivalent.)

DEFINITION 6.3 A function f is analytic on an open set G if for each $a \in G$, f has a power series expansion on some disk with center a.

The thesis of Theorem 5.16 is that if f is defined by a convergent power series on a disk $|z-a| < r$, then f is analytic on that disk. Theorem 3.20 of Chapter 1

SECTION 6. ANALYTIC FUNCTIONS

shows that a rational function is analytic on the plane with the zeros of the denominator removed. The exponential, sine, and cosine are defined for complex z by the series in (2.8) of Chapter 1. Since the series converge for all complex z, the functions are analytic on the whole plane. The identity

$$e^{iz} = \cos z + i \sin z \tag{6.4}$$

is established by inspection of the three series.

THEOREM 6.5 (Unique Continuation) Let f and g be analytic on the connected open set G. If for some point $a \in G$ there is a sequence $a_n \to a$, $a_n \neq a$, with $f(a_n) = g(a_n)$ for all n, then f and g are identical on G.

Proof. It is enough to show that if $f(a_n) = 0$ for all n, then f is identically 0 on G. It is shown first that f is identically 0 on some disk with center a. If not, let a_N be the first nonzero coefficient in the series expansion at a, so that

$$f(z) = \sum_{n=N}^{\infty} a_n(z-a)^n = (z-a)^N \left(a_N + \sum_{n=N+1}^{\infty} a_n(z-a)^{n-N} \right).$$

Call h(z) the term in parentheses. As a convergent power series, h is continuous at a. For each n, $h(a_n) = 0$ because $f(a_n) = 0$ while $(a_n - a)^N \neq 0$.

Therefore, $h(a) = 0$. However, $h(a) = a_N \neq 0$. This means that all the coefficients in the series are 0, so f is 0 on a disk with center a.

Let b be any point in G, and join a to b by a path p in G. By what has been shown, the function $f(p(s))$ is identically 0 on some initial interval $[t_i, t]$. Let $[t_i, \bar{t}]$ be the largest initial interval on which $f(p(s))$ is identically 0, i.e. $\bar{t} = \sup\{t : f(p(s)) = 0 \text{ on } [t_i, t]\}$. Let $\underline{t}$ be the smallest t such that $p(t)$ is constant on $[\underline{t}, \bar{t}]$. (Why can't $\underline{t} = t_i$?) Since $p(s)$ is not identically equal to $p(\bar{t})$ on any interval to the left of $\underline{t}$, there is a sequence t_n increasing to $\underline{t}$ with $c_n = p(t_n) \neq c = p(\bar{t})$, and $f(c_n) = 0$ because $t_n < \underline{t}$. The first part of the proof, applied to the sequence c_n gives that f is identically 0 on a disk with center c, hence that $f(p(s))$ is identically 0 on an interval larger than $[t_i, \bar{t}]$ - unless $\bar{t} = t_f$, in which case $f(b) = 0$ as required.

As a first example of unique continuation, consider the exponential and trigonometric identities

$$e^{z+w} = e^z e^w, \tag{6.6}$$

$$\sin(z+w) = \sin z \cos w + \cos z \sin w \tag{6.7}$$

$$\cos(z+w) = \cos z \cos w - \sin z \sin w. \tag{6.8}$$

If w is fixed and real, the three are known to hold for real z. Since both sides are analytic functions of z, it follows from unique continuation that the identities

SECTION 6. ANALYTIC FUNCTIONS   59

hold for complex z. Now fix z complex. By the above, the identities hold for real w, therefore by unique continuation for complex w.

As a second example, consider the question of defining an analytic logarithm on the right half plane. For any real a > 0, expand $1/z$ about the center a and integrate term by term to obtain a power series L_a that converges on $|z-a| < a$. (The integration constant is chosen so that $L_a(1) = 0$.) By the construction, if b > a, then $L_b(z) = L_a(z)$ for z real and < a. By unique continuation $L_b(z) = L_a(z)$ for z complex, $|z-a| < a$, so the L_a fit together to define an analytic function L on the right half plane. By the construction, $e^{L(z)} = z$ for z real. If the composite of two analytic functions is analytic, as is proved later, unique continuation shows that the identity holds for all z.

7. PREVIEW.

The rest of the notes are devoted to proofs of some of the basic elementary theorems about analytic functions and to examples of their use in computation. The present section contains a preview of some theorems and their uses.

THEOREM 7.1   A function f is analytic on an open set G if and only if it is complex differentiable at each point of G. The radius of convergence for the

series expansion at a point $a \in G$ is at least the distance from a to the complement of G.
This is Theorem 4.5 of Chapter 3.

COROLLARY 7.2 The sum, product, quotient, and composite of analytic functions are analytic, provided, in the case of the quotient, the denominator is not 0.

In many cases the theorem and corollary allow a simple determination of the disk of convergence. Consider, for example, the tangent, defined by

$$\tan z = \sin z / \cos z. \qquad (7.3)$$

The derivatives of the tangent become increasingly complex, and it is not feasible to use Taylor's formula to decide whether the tangent has series expansions, where they converge, how fast they converge, etc. However, since both the sine and cosine are analytic everywhere, the tangent is analytic on the plane with the zeros of the cosine removed. From (6.4) it follows that

$$\cos z = (e^{iz} + e^{-iz})/2 \quad \text{and} \quad \sin z = (e^{iz} - e^{-iz})/2i, \qquad (7.4)$$

therefore, that $\cos z = 0$ if and only if $e^{iz} = -e^{-iz}$. If $z = x+iy$ with x and y real, this gives $e^{-y}e^{ix} = -e^{y}e^{-ix}$. When x is real (6.4) shows that $|e^{ix}| = 1$. Therefore $e^{-y} = e^{y}$, which requires $y = 0$, and therefore z real. The complex zeros of the cosine are the same as the real zeros, namely the odd multiples of $\pi/2$,

SECTION 7. PREVIEW 61

and the tangent is analytic on the plane with the odd multiples of $\pi/2$ removed. With any center, the radius of convergence for the series expansion is the distance from that center to the nearest odd multiple of $\pi/2$.

Even though the radius of convergence with a given center is now known the tangent still presents a typical computational problem. Any given coefficient a_n can be determined from the previous ones by formula (4.13) of Chapter 1, but the a_n do not form a simple pattern that allows error evaluations. These can be achieved with Cauchy's Inequalities, which are stated here and also as Theorem 4.9 of Chapter 3.

THEOREM 7.5 (Cauchy's Inequalities) If f is analytic on the disk $|z-a| < r$ and satisfies $|f(z)| \leq M$ on $|z-a| < r$, then the coefficients in the series for f at a satisfy $|a_n| \leq M/r^n$.

Example 7.6 Compute, with error $\leq 10^{-6}$,

$$\int_0^{\pi/8} \sqrt{x} \, \tan x \, dx \, . \quad \text{If}$$

$$\tan z = \sum_{n=0}^{\infty} a_n z^n, \quad \text{then} \quad \int_0^y \sqrt{x} \, \tan x \, dx = \sum_{n=0}^{\infty} \frac{a_n y^{n+3/2}}{n+3/2},$$

so in stopping with $n = N$, the error is at most

$$\sum_{n=N+1}^{\infty} |a_n| |y|^{n+3/2}/(n+3/2).$$

Suppose that $|\tan z| \leq M$ on $|z| \leq 3\pi/8$. Cauchy's Inequalities give $|a_n| = M/(3\pi/8)^n$. This implies that for $|y| \leq \pi/8$, the error is at most

$$\frac{M(\pi/8)^{3/2}}{N+5/2} \sum_{n=N+1}^{\infty} (1/3)^n = \frac{M(\pi/8)^{3/2}}{(2N+5)3^N},$$

the last equality coming from summing the geometric series. It is shown in Problem 2, Section 4, Chapter 3 that $M \leq 3.85$. Put this in the last estimate. A bit of arithmetic shows that the error is $\leq 10^{-6}$ for $N = 10$. Finally, the coefficients $a_0, \ldots, a_{10}$ are calculated by the formula

$$T_0^{10}(\tan z) T_0^{10}(\cos z) = T_0^{10}(\sin z) \mod(z^{11})$$

of section 4, Chapter 1. Since the tangent is an odd function, it is known a priori that all even coefficients are 0, so the only ones needing calculation are a_1, a_3, a_5, a_7, and a_9, and the 10's above can be replaced by 9's in the case of the tangent and sine, and by 8 in the case of the cosine.

In general, if $|f(z)| \leq M$ on $|z-a| < r$, so that $|a_n| \leq M/r^n$, then on $|z-a| \leq r_1 < r$ there is the error evaluation

$$\sum_{n=N+1}^{\infty} |a_n| |z-a|^n \leq M \sum_{n=N+1}^{\infty} (r_1/r)^n = \frac{M(r_1/r)^{N+1}}{1-(r_1/r)} \quad (7.7)$$

PROBLEMS

SECTION 2.

If z and w are complex vectors, the inner product is defined by $<z,w> = \sum_{j=1}^{n} z_j \bar{w}_j$, hence $|z| = \sqrt{|<z,z>|}$.

1. Prove the Cauchy-Schwarz inequality:

$|<z,w>| \leq |z||w|$, via

a) For real t, the function

$f(t) = |z-tw|^2 = |z|^2 + 2t\text{Re}(<z,w>) + t^2|w|^2$

is real and non-negative.

b) Let t_0 be the point where $f(t)$ is minimum. $f(t_0) \geq 0$ implies $\text{Re}(<z,w>) \leq |z||w|$.

c) Replace z by cz with $|c| = 1$ and $<cz,w> = |<z,w>|$.

2. Prove Minkowski's inequality:

$|z+w| \leq |z|+|w|$.

Hint: Expand $|z+w|^2$ and use Cauchy-Schwarz.

SECTION 3.

1. Give the definition of convergence for series of complex vectors.

2. Prove that

$$\left|\sum_{k=K}^{\infty} a_k\right| \leq \sum_{k=K}^{\infty} |a_k|, \quad a_k \text{ complex vectors.}$$

3. Prove the theorems of the section (except Theorem 3.20) for series of complex vectors.

4. Let $p(w) = w^m + \sum_{k=0}^{m-1} a_k w^k$. If $|a_k| \leq M$, then

$$|w|-M-1 \leq |p(w)/w^{m-1}| \leq |w|+M+1 \quad \text{for} \quad |w| \geq M+1$$

5. Find N so that $\sum_{n=N}^{\infty} |a_n| \leq 10^{-6}$, where

$$a_n = \frac{n^5+5n^3-1}{n^9-7n^6-3n+1}, \quad a_n = \frac{3n^2+\sin n}{n^5-\cos n}$$

SECTION 4.

1. Show that the solution y to the differential equation (4.1) satisfies

$$|y(x)| \leq |y_0| + \frac{M_0}{M}(e^{M(b-a)}-1)$$

2. Convert the differential equation

$$u^n = \sum_{k=0}^{n-1} a_k(x) u^k + f$$

to vector form and use problem 1 to find a bound for

$|u(x)|$ in terms of

$$\|a\|^2 = \max\ 1+ \sum_{k=0}^{n-1} |a_k(x)|^2\ ,$$

$\|f\| = \max |f(x)|$, and $u(x_o), \ldots u^{n-1}(x_o)$.

SECTION 6.

1. The proof of (6.6-8) used the facts that e^{z+b}, $\sin(z+b)$, and $\cos(z+b)$ are analytic in z when b is fixed. Use Theorem 5.16 to prove this.

2. More generally, let

$$f(z) = \sum_{n=0}^{\infty} a_n(z-a)^n \quad \text{on } |z-a| < r.$$

Show that if $|b-a| < r$, then $f(z+b)$ is analytic on $|z| < r-|b-a|$.

Hint: Expand f around b.

3. Prove the identity
$\sin z^2 + \cos z^2 = 1$ for z complex.

4. Find the complex z for which:
$\sin(z) = 0$, $\sin(z) = 1$, $\sin(z) = -1$, $\cos(z) = 0$,
$\cos(z) = 1$, $\cos(z) = -1$.
Hint: For $\cos(z) = 0$, see the next section.

5. More generally, given a, find the z for which
$\sin(z) = \sin(a)$; also for which $\cos(z) = \cos(a)$.

Also the z for which $e^z = e^a$.

SECTION 7.

1. Find the points where the functions in Problem 1, Section 4, Chapter 1 are not differentiable.

2. Use Theorem 7.1 to find the radius of convergence of the series at a given point a for the functions in Problem 1, Section 4, Chapter 1.

3. Use the Taylor series at 0 to show that

$$|\sin(z)| \leq e^{|z|}-1, \quad |\cos(z)| \leq e^{|z|}, \quad |e^z| \leq e^{|z|}.$$

4. Use the Taylor series at 0 to show that

$$|1-\cos(z)| \leq \frac{|z|^2}{2} e^{|z|}$$

5. Use Cauchy's Inequalities to find N so that $|T_0^N f(x) - f(x)| \leq 10^{-6}$ for $|x| \leq 1/2$ for the functions in in Problem 1, Section 4, Chapter 1.

6. Let $w = \sqrt{x^2+1}$ on $|x| < 1$, assuming that $w^2 = z^2+1$ has an analytic solution on $|z| < 1$. Use Cauchy's Inequalities to find N so that $|w - T_0^N w| \leq 10^{-6}$ on $|x| \leq 1/2$.

7. Show that if $|a_k| \leq M$, and w satisfies the equation

$$w^m + \sum_{k=0}^{m-1} a_k w^k = 0,$$

then $|w| \leq M+1$. (See Problem 4 of Section 3.)

8. Try to establish a bound for the function u in Problem 2, Section 4, but for complex x (for use in Cauchy's inequalities). (See Chapter 6.)

3. POWER SERIES AND COMPLEX DIFFERENTIABILITY

1. PATHS IN THE COMPLEX PLANE C.

DEFINITION 1.1 A path in a set $G \subset C$ is a continuous function p defined on a closed interval $[t_i, t_f]$ with values in G. The points $a = p(t_i)$ and $b = p(t_f)$ are the initial and final points of the path, and the path is said to join a and b, or to go from a to b. The path p^{-1} is defined by

$$p^{-1}(t) = p(t_f + t_i - t) . \qquad (1.2)$$

If q is another path, defined on $[s_i, s_f]$, whose initial point coincides with the final point of p, the path pq is defined by

$$(pq)(t) = \begin{cases} p(t) & \text{for } t_i \leq t \leq t_f \\ q(t - t_f + s_i) & \text{for } t_f \leq t \leq t_f + s_f - s_i. \end{cases} \qquad (1.3)$$

p^{-1} is a path from b to a, and pq is a path from the initial point of p to the final point of q.

The main paths needed in the sequel are the following.

If a and b are points in C, the segment [a,b] is the path

$$[a,b](t) = a + t(b-a), \quad 0 \leq t \leq 1 . \qquad (1.4)$$

If T is a (two dimensional) triangle with vertices a, b, and c ordered so that T is on the left on each

SECTION 1. PATHS IN THE COMPLEX PLANE C

segment, then ∂T, the boundary of T, is the path

$$\partial T = [a,b][b,c][c,a]. \tag{1.5}$$

If D is the disk with center a and radius r, then ∂D is the path

$$\partial D(t) = a + re^{it}, \quad 0 \leq t \leq 2\pi. \tag{1.6}$$

An arc of ∂D is the restriction of ∂D to a subinterval $[t_i, t_f]$.

Let D_a be a disk with center a inside a disk D. The horizontal and vertical lines through a divide ∂D and ∂D_a into pairs of corresponding arcs, and the region $D - D_a$ into four corresponding regions. Let p and p_a be corresponding arcs, with G the corresponding region, and with p and p_a defined on intervals $[t_i, t_f]$ and $[s_i, s_f]$. Then ∂G is the path

$$\partial G = p[p(t_f), p_a(s_f)] p^{-1} [p_a(s_i), p(t_i)]. \tag{1.7}$$

A picture of these last paths is shown at the beginning of section 4. Strictly speaking, pictures are not legitimate, for the path p is not the set
$P = \{p(t): t_i \leq t \leq t_f\}$ = the range of p = the set of points on p. For instance, p and p^{-1} have the same range, but are not the same path.

The function p determines not only the range P, but also a "velocity of motion" along P (namely the derivative p'). This velocity involves direction, p'(t)/ p'(t) , and speed, $|p'(t)|$. In the present context the direction is important, but the speed is not, as is indicated by Theorem 2.4 in the next section. For this reason, with simple paths like those above, a picture showing the range and direction does provide an adequate representation of the path itself. (It is important, nevertheless, to be able to translate such pictures into formulas.)

2. PATHS INTEGRALS.

DEFINITION 2.1 The path p defined on $[t_i, t_f]$ is smooth if the derivative p' exists and is uniformly continuous on the open interval. p is piecewise smooth if $[t_i, t_f]$ can be partitioned into a finite number of sub-intervals on which p is smooth. The length of a piecewise smooth path p is given by

$$L(p) = \int_{t_i}^{t_f} |p'(t)| \, dt$$

For the present, all paths are assumed piecewise smooth. It is clear that if p and q are piecewise smooth, then so are p^{-1} and pq; and it is clear that the special paths at the end of the last section are piecewise smooth. When the interval on which a path is defined is not specified, it is assumed to be $[t_i, t_f]$.

DEFINITION 2.2 If the function f is continuous on the range of the path p, the integral of f along p is defined by

$$\int_p f(z)dz = \int_{t_i}^{t_f} f(p(t))p'(t)dt \qquad (2.3)$$

It is always assumed that the function f appearing in a path integral is continuous on the range of the path.

THEOREM 2.4 If $a: [s_i, s_f] \to [t_i, t_f]$ (onto) is piecewise smooth and strictly increasing, then $p \circ a$ is piecewise smooth and

$$\int_p f(z)dz = \int_{p \circ a} f(z)dz, \quad L(p) = L(p \circ a) \qquad (2.5)$$

If a is strictly decreasing, then

$$\int_p f(z)dz = - \int_{p \circ a} f(z)dz, \quad L(p) = L(p \circ a) \qquad (2.6)$$

Proof. Make the change of variable $t = a(s)$ in the integral (2.3).

The following are obvious from Theorem 2.4.

$$\int_{p^{-1}} f(z)dz = - \int_p f(z)dz, \qquad (2.7)$$

$$\int_{pq} f(z)dz = \int_p f(z)dz + \int_q f(z)dz \qquad (2.8)$$

LEMMA 2.9 If $|f(z)| \leq M$ on the range of p, then

$$\left|\int_p f(z)dz\right| \leq ML(p).$$

THEOREM 2.10 If $f(z) = F'(z)$ for each z in the range of p, then

$$\int_p f(z)dz = F(p(t_f)) - F(p(t_i)). \qquad (2.11)$$

Proof. If $G(t) = F(p(t))$, then $G'(t) = F'(p(t))p'(t)$, so the integrand on the right of (2.3) is G'.

When p is a segment [a,b] it is customary to write

$$\int_a^b f(z)dz \quad \text{in place of} \quad \int_{[a,b]} f(z)dz.$$

3. CAUCHY'S INTEGRAL THEOREM.

DEFINITION 3.1 A set G is convex if for each pair of points a and b in G, the segment [a,b] lies in G.

A simplified version of Cauchy's Integral Theorem is:

THEOREM 3.2 (Cauchy's Integral Theorem) If f is complex differentiable on a convex open set G, the integral of f along every closed path in G is 0.

SECTION 3. CAUCHY'S INTEGRAL THEOREM

The proof is divided into three steps, the first coming from Theorem 2.10, which gives

THEOREM 3.3 If f has a complex primitive on an open set G, then the integral of f along any closed path in G is 0.

The second and third steps are:

THEOREM 3.4 Let G be a convex open set. If the integral over ∂T is 0 for each triangle T with vertices in G, then f has a complex primitive on G.

THEOREM 3.5 Let G be a convex open set. If f is complex differentiable on G, then the integral over ∂T is 0 for each triangle T with vertices in G.

Proof of Theorem 3.4. Fix $c \in G$ and define

$$F(z) = \int_c^z f(w)dw \qquad (3.6)$$

According to Problem 2, Section 2,

$$F(a+h) - F(a) = \int_a^{a+h} f(z)dz. \qquad (3.7)$$

By definition, $[a, a+h](t) = a+th$, so for any constant c

$$\int_a^{a+h} c\, dz = \int_0^1 ch\, dt = ch . \qquad (3.8)$$

With c = f(a), (3.7) and (3.8) give

$$\left|\frac{F(a+h)-F(a)}{h} - f(a)\right| = \left|\frac{1}{h}\int_a^{a+h} f(z)-f(a)\ dw\right|. \qquad (3.9)$$

If $\epsilon > 0$ is given, use the continuity of f at a to find $\delta > 0$ so that if $|z-a| < \delta$, then $|f(z)-f(a)| \leq \epsilon$. If $|h| < \delta$, then each point of [a,a+h] is within distance δ of a, so $|f(z)-f(a)| \leq \epsilon$ on [a,a+h]. By Lemma 2.9 this implies that the right side of (3.9) is $\leq \epsilon$, so that F'(a) = f(a).

Proof of Theorem 3.5. If T is a triangle with vertices in G, divide T into four congruent triangles T_i' by joining the midpoints of the sides of T. Then (picture page 76),

$$\int_{\partial T} f(z)dz = \int_{\partial T_1'} f(z)dz + \int_{\partial T_2'} f(z)dz + \int_{\partial T_3'} f(z)dz + \int_{\partial T_4'} f(z)dz.$$

The integrals over the interior boundary segments cancel because of the definition of the boundary of a triangle. This implies that for at least one of the T_i', call it T_1, it must be true that

$$\left|\int_{\partial T} f(z)dz\right| \leq 4 \left|\int_{\partial T_1} f(z)dz\right|.$$

SECTION 3. CAUCHY'S INTEGRAL THEOREM

Now divide T_1 in the same way to get $T_2 \subset T_1$ with

$$\left| \int_{\partial T} f(z)da \right| \leq 4 \left| \int_{\partial T_1} f(z)dz \right| \leq 4^2 \left| \int_{\partial T_2} f(z)dz \right|.$$

In general, divide T_{n-1} to get $T_n \subset T_{n-1}$ with

$$\left| \int_{\partial T} f(z)dz \right| \leq 4^n \left| \int_{\partial T_n} f(z)dz \right|, \qquad (3.10)$$

$$\text{diam}(T_n) = (1/2)\text{diam}(T_{n-1}) = (1/2^n)\text{diam}(T), \text{ and} \qquad (3.11)$$

$$L(\partial T_n) = (1/2)L(\partial T_{n-1}) = (1/2^n)L(\partial T), \qquad (3.12)$$

where

$$\text{diam}(A) = \sup \{|x-y| : x, y \in A\}. \qquad (3.13)$$

Because of (3.11) there is a point a common to all the triangles T_n: If $a_n \in T_n$, the sequence $\{a_n\}$ is Cauchy, so has a limit a. The proof is finished by using the differentiability at a.

Let $\epsilon > 0$ be given, and choose $\delta > 0$ so that if $|z-a| < \delta$, then

$$f(z) = f(a) + f'(a)(z-a) + h(z) \text{ with } |h(z)| \leq \epsilon |z-a|.$$

Then choose n so that $\text{diam}(T_n) < \delta$. Since both the constant $f'(a)$ and the function $f'(z)(z-a)$ have complex primitives, it follows that

$$\left| \int_{\partial T_n} f(z)dz \right| =$$

$$\left| \int_{\partial T_n} h(z)dz \right| \leq \epsilon(1/2^n)\text{diam}(T)(1/2^n)L(\partial T)$$

and consequently, from (3.10), that

$$\left| \int_{\partial T} f(z)dz \right| \leq \epsilon \, \text{diam}(T)L(\partial T),$$

and therefore that the integral is 0.

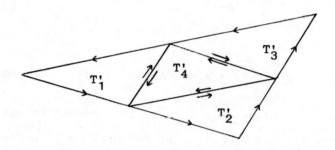

4. CAUCHY'S INTEGRAL FORMULA AND CAUCHY'S INEQUALITIES

THEOREM 4.1 (Cauchy's Integral Formula) If f is complex differentiable on a neighborhood of a closed disk D, and a is inside D, then

$$f(a) = \frac{1}{2\pi i} \int_{\partial D} \frac{f(z)}{z-a} dz \qquad (4.2)$$

SECTION 4. CAUCHY'S FORMULA AND INEQUALITIES

Proof. Let D_a be a disk with center a contained in D. The horizontal and vertical lines through a divide $D-D_a$ into four regions G_i, as shown.

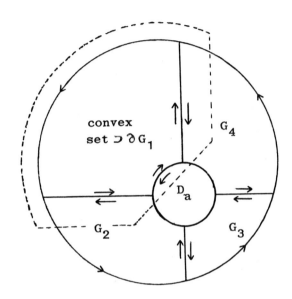

The function $g(z) = (f(z)-f(a))/(z-a)$ is complex differentiable on a convex open set containing ∂G_i, as shown, so the integral of g over ∂G_i is 0. Therefore,

$$0 = \sum_{i=1}^{4} \int_{\partial G_i} g(z)dz = \int_{\partial D} g(z)dz - \int_{\partial D_a} g(z)dz . \quad (4.3)$$

If the radius r of D_a is sufficiently small, then $|g(z)| \leq |f'(a)|+1$ on ∂D_a, so that by Lemma 2.9

78 CHAPTER 3. POWER SERIES AND COMPLEX DIFF.

$$\left|\int_{\partial D} g(z)dz\right| = \left|\int_{\partial D_a} g(z)dz\right| \leq (|f'(a)|+1)\, 2\pi r.$$

Since this holds for every small r, the integral of g over ∂D is 0, and

$$\int_{\partial D} \frac{f(z)}{z-a}\, dz = f(a) \int_{\partial D} \frac{1}{z-a}\, dz .$$

To evaluate the integral on the right, apply (4.3) to $g(z) = 1/(z-a)$ to get

$$\int_{\partial D} \frac{1}{z-a}\, dz = \int_{\partial D_a} \frac{1}{z-a}\, dz = \int_0^{2\pi} \frac{ire^{it}}{re^{it}}\, dt = 2\pi i.$$

To emphasize the fact that a is arbitrary in Cauchy's formula, it is convenient to replace the dummy variable z by w, then a by z, to get

THEOREM 4.4 (Cauchy's Integral Formula) If f is complex differentiable on a neighborhood of a closed disk D, and z is inside D, then

$$f(z) = \frac{1}{2\pi i} \int_{\partial D} \frac{f(w)}{w-z}\, dw$$

THEOREM 4.5 (Series Expansion) Let f be complex differentiable on the open set G, let a be a point in G, and let r_a be the distance from a to the boundary of G. Then f has a series expansion on the disk $|z-a| < r_a$,

SECTION 4. CAUCHY'S FORMULA AND INEQUALITIES

$$f(z) = \sum_{n=0}^{\infty} a_n(z-a)^n, \text{ with } a_n = \frac{1}{2\pi i} \int_{\partial D} \frac{f(w)}{(w-a)^{n+1}} dw \quad (4.6)$$

where D is any closed disk in G with a in the interior. This implies that $f^n(a) = n!a_n$, and that the Taylor polynomial $T_a^m f$ is characterized by the equation

$$f = T_a^m f \mod (z-a)^{m+1}.$$

Proof. Fix z with $|z-a| < r_a$, then fix r with $|z-a| < r < r_a$. The proof of Cauchy's formula shows that the integral on the right of (4.6) is independent of D, so take D to be the disk $|z-a| \leq r$. Using this D, expand $1/(w-z)$ in (4.4) in the geometric series

$$\frac{1}{w-z} = \frac{1}{w-a} \sum_{n=0}^{\infty} \frac{(z-a)^n}{(w-a)^n} \quad \text{for } \frac{|z-a|}{|w-a|} < 1. \quad (4.7)$$

When $w \in \partial D$, $|w-a| = r > |z-a|$, so the series in (4.7) converges uniformly in w, for $w \in \partial D$. By Theorem 4.9 of Chapter 2, it can be inserted in (4.4) and integrated term by term to yield (4.6).

Theorem 4.5 asserts that if f is complex differentiable on G, then f is analytic on G and shows that the expansion around the center a is valid on the largest disk with center a and contained in G. Therefore, f is analytic on G if and only if f is complex differentiable on G.

THEOREM 4.8 (Cauchy's Inequalities) If f is complex differentiable on the disk $|z-a| < r$, and satisfies $|f(z)| \leq M$ on this disk, then

$$|a_n| \leq M/r^n \tag{4.9}$$

Proof. Let D be the disk with center a and radius $r_o < r$. On ∂D, the integrand in (4.6) is at most M/r_o^{n+1}, so by Lemma 2.9

$$|a_n| \leq \frac{M \, 2\pi \, r_o}{2\pi \, r_o^{n+1}} = M/r_o^n$$

Since this holds for all $r_o < r$, it also holds for r.

This pins down the M, the existence of which was proved earlier in Lemma 5.4 of Chapter 2.

THEOREM 4.10 (Maximum Modulus Theorem) Let f be analytic on the connected open set G. If $|f(z)|$ has a local maximum, then f is constant on G.

Proof. Suppose $|f|$ has the local maximum M at the point a. Multiply f by a constant, to obtain $f(a) = M$. Let D be the disk with center a and radius r small enough so that $|f(z)| \leq M$ on D. If $f(z) = u(z)+iv(z)$ with u and v real, Cauchy's formula gives

$$M = f(a) = \frac{1}{2\pi i} \int_{\partial D} \frac{f(z)}{z-a} dz = \frac{1}{2\pi i} \int_0^{2\pi} f(a+re^{it}) dt$$

$$= \frac{1}{2\pi} \int_0^{2\pi} u(a+re^{it}) dt,$$

the integral of v being 0 because the left side is real. Since $u(a+re^{it})$ is real and continuous and $\leq M$ everywhere, it follows that $u(a+re^{it}) = M$ everywhere, therefore that $v(a+re^{it}) = 0$ everywhere. Consequently, $f = M$ on ∂D, and Unique Continuation requires that $f = M$ everywhere on G.

THEOREM 4.11 Let f be analytic on the disk $|z-a| < r$ and continuous on the closed disk. If $|f(z)| \leq M$ on $|z-a| = r$, then

$$|f(z) - T_a^N f(z)| \leq M \left(\frac{r_1}{r}\right)^{N+1} \frac{r}{r-r_1} \quad \text{on } |z-a| \leq r_1 < r. \tag{4.12}$$

Proof. If the maximum of $|f(z)|$ on the closed disk were at an interior point, then by the Maximum Modulus Theorem f would have to be constant on the disk. Therefore $|f(z)| \leq M$ on $|z-a| \leq r$, and Cauchy's inequalities give

$$|f(z) - T_a^N f(z)| = \left|\sum_{n=N+1}^{\infty} a_n (z-a)^n\right| \leq M \sum_{n=N+1}^{\infty} (r_1/r)^n .$$

82   CHAPTER 3. POWER SERIES AND COMPLEX DIFF.

In using the theorem for computation it is necessary to fix an r less than or equal (usually less than) the radius of convergence, but larger than $|z-a|$ for all z that enter the computation; then to find an M. Even though M increases with r, usually it is best to take r nearly as large as possible because of the geometric dependence of the error on r.

Example 4.13 Calculate, with error $\leq 10^{-6}$,

$$\int_0^{\pi/2} f(x)dx \quad \text{for } f(x) = (\pi/2-x)\tan x.$$

Since $f(z)$ is unbounded near $z = -\pi/2$, 0 is not suitable series center. The center is taken at $\pi/4$, so that the expansion is

$$f(z) = \sum_{n=0}^{\infty} a_n (z-\pi/4)^n .$$

Since $\cos(z) = \sin(\pi/2-z)$, $\cos(z) = (\pi/2-z)h(z)$, where h is analytic everywhere and is 0 at the odd multiples of $\pi/2$, except for $\pi/2$ itself. It follows that $(\pi/2-z)\tan z$ is analytic on $|z-\pi/2)| < \pi$ and that

$$\max_{|z-\pi/2|=r} |(\pi/2-z)\tan z| = \max_{|z-\pi/2|=r} |(\pi/2-z)\cot(\pi/2-z)| \leq$$

$$r \max_{|w|=r} |\cot w| .$$

SECTION 4. CAUCHY'S FORMULA AND INEQUALITIES

According to Problem 2 for this section, this is at most 56.35 for $r = 7\pi/8$, so

$$|f(z)| = |\pi/2-z||\tan z| \leq 56.35 \quad \text{on} \quad |z-\pi/2)| \leq 7\pi/8.$$

Since $|z-\pi/2| \leq |z-\pi/4| + \pi/4$,

$$|f(z)| \leq 56.35 \quad \text{on} \quad |z-\pi/4| \leq 5\pi/8$$

and the coefficients with this center satisfy

$$|a_n| \leq 56.35/(5\pi/8)^n .$$

Since

$$\int_{\pi/4-y}^{\pi/4+y} f(x)dx = 2 \sum_{n=0}^{\infty} \frac{a_{2n}}{2n+1} y^{2n+1} ,$$

with $y = \pi/4$, the error in stopping at $n = N$ is at most

$$(\pi/4) \times 2 \times 56.35/(2N+3) \sum_{n=N+1}^{\infty} (2/5)^{2n} =$$

$$42.16 \times (2/5)^{2N+1}$$

This is less than 10^{-6} for $N = 8$, so coefficients $a_0, \ldots, a_{16}$ are needed.

PROBLEMS

SECTION 2.

1. Referring to the picture in Section 3, use Definition 1.5 and Theorems 2.4 and 2.8 to show that

$$\int_{\partial T} f(z)dz = \sum_{i=1}^{4} \int_{\partial T_i'} f(z)dz$$

2. Referring to the picture in Section 4, use Definition 1.7 and Theorems 2.4 and 2.8 to show that

$$\int_{\partial D} f(z)dz - \int_{\partial D_a} f(z)dz = \sum_{i=1}^{4} \int_{\partial G_i} f(z)dz$$

3. Let $F(w) = \int_c^w f(z)dz$. Show that

$$F(a+h) - F(a) = \int_a^{a+h} f(z)dz \pm \int_{\partial T} f(z)dz$$

where T is the triangle with vertices c, a, and a+h. On what does the $\pm$ sign depend?

4. If f has a complex primitive on the range of p, then

$$\int_p f(z)dz = 0.$$

CHAPTER 3. POWER SERIES AND COMP. DIFF. - PROBLEMS

SECTION 3.

1. Referring to the picture in Section 4, let f be complex differentiable on a neighborhood of the closed disk D, and let $g(z) = \frac{f(z)-f(a)}{z-a}$.

Show that $\int_{\partial G_i} g(z)dz = 0$.

SECTION 4.

1. Use Lagrange multipliers to show that the max and min of $|\sin z|$ and $|\cos z|$ on $|z| = r$ occur and $x = 0$, $y = \pm r$ or at $y = 0$, $x = \pm r$.

2. Conclude that

$\max_{|z|=r} |\sin z| = \frac{1}{2} (e^{2r}+e^{-2r}-2)^{1/2}$

$\min_{|z|=r} |\sin z| = |\sin r|$

$\max_{|z|=r} |\cos z| = \frac{1}{2} (e^{2r}+e^{-2r}+2)^{1/2}$

$\min_{|z|=r} |\cos z| = |\cos r|$.

3. Conclude that

$\max_{|z| \leq r} |\sin z| = \frac{1}{2} (e^{2r}+e^{-2r}-2)^{1/2}$

$\max_{|z| \leq r} |\cos z| = \frac{1}{2} (e^{2r}+e^{-2r}+2)^{1/2}$

4. What about $\min_{|z| \leq r} |\sin z|$ and $\min_{|z| \leq r} |\cos z|$?

5. In general, what about $\min_{|z| \leq r} |f(z)|$ vs. $\min_{|z| = r} |f(z)|$ when

f is analytic on $|z-a| < r$ and continuous on $|z-a| \leq r$.

6. Discuss the role of the Maximum Modulus Theorem in Example 4.13.

7. Calculate $\displaystyle\int_0^1 \frac{\sin x}{e^x - 1}\, dx$ with error $\leq 10^{-6}$.

CHAPTER 4.   LOCAL ANALYTIC FUNCTIONS

1. LOGARITHMS.

A logarithm function has two features. On the one hand, it is a primitive of $1/z$. On the other it is an inverse to the exponential. However, on the one hand $1/z$ does not have a primitive on any ring around 0 (because its integral over a circle around 0 is not 0), and on the other, the exponential is not one to one. If $w = u+iv$, then

$$e^w = e^{u+iv} = e^u e^{iv} = e^u(\cos v + i\sin v).$$

If u and v are real, then e^u is real and positive, and $|\cos v + i\sin v| = 1$. Consequently,

$e^w = z$ if and only if $|z| = e^u$ and $z/|z| = \cos v + i\sin v.$ \hfill (1.1)

DEFINITION 1.2   If $z \neq 0$, an argument of z is any number v satisfying $z/|z| = \cos v + i\sin v$. Any two arguments of z differ by an integer multiple of 2π. The principal argument is the one satisfying
$-\pi < \mathrm{Arg}(z) \leq \pi.$

THEOREM 1.3 w is a logarithm of z (i.e. $e^w = z$) if and only if

$\mathrm{Re}(w) = \log|z|$ and $\mathrm{Im}(w)$ is an argument of z.

Consequently, any two logarithms of z differ by an integer multiple of $2\pi i$.

DEFINITION 1.4 A branch of the logarithm is a continuous function L defined on a connected open set G and satisfying

$$e^{L(z)} = z \text{ for all } z \in G.$$

THEOREM 1.5 Let L be a branch of the logarithm on the connected open set G. For any integer n, the function $M(z) = L(z)+2n\pi i$ is also a branch of the logarithm on G. Conversely, if M is any branch of the logarithm on G, then there is an integer n such that $M(z) = L(z)+2n\pi i$ for all $z \in G$.

Proof. The first part is immediate. The second is a consequence of the following lemma, since the function $F(z) = (M(z)-L(z))/2\pi i$ takes only integer values.

LEMMA 1.6 If the continuous real valued function F on the connected open set G takes the values c and d, then it takes every value between.

Proof. Let p be a path joining the arbitrary points a and b, and set $P(t) = F(p(t))$. By the Intermediate Value Theorem, P takes every value between $f(a) = P(t_i)$ and $F(b) = P(t_f)$.

Although branches of the logarithm do not exist on any ring around 0, they do exist on the plane with any ray removed. For $a \neq 0$, let G_a be the plane with the ray composed of non-positive multiples of a removed.

THEOREM 1.7 The function

$$L_a(z) = \int_a^z (1/w)dw + \log|a| + i\text{Arg}(a) \qquad (1.8)$$

is a branch of the logarithm on G_a.

Proof. For any $z \in G_a$, there is a convex open $G \subset G_a$ that contains both a and z. It follows from Theorems 3.4 and 3.5 of Chapter 3 that L_a is a primitive of $1/z$ on G_a, hence also that L_a is analytic on G_a. For z real and positive, $L_1(z)$ is the usual real logarithm, so that

$$e^{L_1(z)} = z$$

for z real and positive, therefore for all $z \in G_1$, by unique continuation. Thus L_1 is a branch of the logarithm on G_1.

Since $L_1(z)$ is a logarithm of z, it must have the form

$$L_1(z) = \log|z| + i\text{Arg}(z) + 2n_z \pi i,$$

where n_z is an integer. Since the other summands are

continuous on G_1, n_z must be continuous too, therefore constant, while the value at 1 is 0, so

$$L_1(z) = \log|z| + i\mathrm{Arg}(z). \tag{1.9}$$

For arbitrary a, the set $G_a \cap G_1$ where both L_a and L_1 are defined is the union of two connected open sets H_1 and H_2. Suppose first that a is not a negative real number, and let H_1 be the one containing a. Since L_a and L_1 are both primitives of $1/z$, they differ by a constant on H_1, and (1.9) shows that they are equal at a. Therefore, $L_a = L_1$ is a branch of the logarithm on H_1, and by unique continuation it is a branch of the logarithm on G_a.

Now let a be real and negative. Since the segment [a,ai] and the quarter circle C joining a and ai are contained in a convex set on which $1/z$ is analytic, Cauchy's theorem gives

$$\int_a^{ai} (1/w)dw = -\int_C (1/w)dw = -\int_{\pi/2}^{\pi} i|a|e^{it}/|a|e^{it}\, dt = -\pi i/2$$

so that $L_a(ai) = -\pi i/2 + \log|a| + \pi i$. According to (1.9) this is also the value of $L_1(ai)$. As above, this implies that $L_a = L_1$ on the upper half plane, hence that L_a is a branch of the logarithm on G_a.

COROLLARY 1.10 If L is a branch of the logarithm on the connected open set G, then L is analytic on G. (i.e. a continuous solution to the equation $e^{L(z)} = z$ is automatically analytic).

Proof. Theorem 1.5 shows that if one branch is analytic, then all branches are, and it has just been shown that there exist analytic branches on a neighborhood of any point except 0.

This is an example of a general result to the effect that if f is analytic, g is continuous, and $f(g(z)) = z$, then g is analytic, a result that is essentially proved in the next section.

2. LOCAL SOLUTIONS TO ANALYTIC EQUATIONS.

DEFINITION 2.1 Let G_z and G_w be connected open sets. A function F on $G_z \times G_w$ is analytic if for each $z_0 \in G_z$, $F(z_0, w)$ is analytic on G_w, and for each $w_0 \in G_w$, $F(z, w_0)$ is analytic on G_z. A remarkable theorem of Hartogs guarantees that F, $\partial F/\partial z$, and $\partial F/\partial w$ are continuous on $G_z \times G_w$.

The topic of this section is the local solution of the analytic equation $F(z,w) = 0$. One example is the equation $f(w) - z = 0$ for the inverse of the function f. The branches of the logarithm defined in the last section are local solutions to $e^w - z = 0$. Other common equations take the form

$$w^m + \sum_{K=0}^{m-1} a_k(z)w^k = 0. \tag{2.2}$$

The solutions are called algebroid functions when the a_k are analytic on $G = G_z$, and algebraic functions when the a_k are polynomials. The n-th root for example, satisfies $w^n - z = 0$. The main theorem on local solutions is as follows. G_z and G_w are open and connected.

THEOREM 2.3 Let F be analytic on $G_z \times G_w$, and let there be no point (z,w) with both $F(z,w) = 0$ and $(\partial F/\partial w)(z,w) = 0$. If $F(a,b) = 0$, there is one and only one function w_{ab} which is analytic on a disk with center a and satisfies $F(z, w_{ab}(z)) = 0$ and $w_{ab}(a) = b$. If K_z and K_w are compact sets in G_z and G_w, there is a number $r > 0$ such that the radii of convergence of the w_{ab} are $\geq r$ for $a \in K_z$, $b \in K_w$, and $F(a,b) = 0$.

The proof of Theorem 2.3 is based partly on a "fixed point" theorem.

THEOREM 2.4 (Contraction Fixed Point Theorem) Let D be a closed set and let F be a function from D to itself. If there is a number $M < 1$ such that

$$|F(w_2) - F(w_1)| \leq M|w_2 - w_1|,$$

there is a unique point $w \in D$ such that $F(w) = w$. (For, any M, the condition on F is called a "Lipschitz condition; when $M < 1$, F is a "contraction".)

SECTION 2. LOCAL SOLUTIONS TO ANALYTIC EQUATIONS 93

Proof. As in the proof of Picard's theorem, let w_0 be any point in D and let $w_n = F(w_{n-1})$. If $w_n \to w$, then $F(w_{n-1}) \to w$, while also $F(w_{n-1}) \to F(w)$, as F is continuous. Therefore, $F(w) = w$. (w must lie in D because D is closed.) It is apparent that there cannot be more than one fixed point, for if w_1 and w_2 are fixed, then $|w_2-w_1| = |F(w_2)-F(w_1)| \leq M|w_2-w_1|$, with $M < 1$.

To prove the convergence, note that, as in Picard's theorem,

$$|w_2-w_1| = |F(w_1)-F(w_0)| \leq M|w_1-w_0|$$

$$|w_3-w_2| = |F(w_2)-F(w_1)| \leq M|w_2-w_1| \leq M^2|w_1-w_0|$$

so that in general

$$|w_{k+1}-w_k| \leq M^k|w_1-w_0|.$$

If $n > m$, then

$$|w_n-w_m| \leq \sum_{k=m}^{n-1}|w_{k+1}-w_k| \leq |w_1-w_0|\sum_{k=m}^{\infty}M^k = |w_1-w_0|\frac{M^m}{1-M}.$$

so the sequence $\{w_n\}$ is Cauchy.

The main theorem is proved first for the equation $f(w) = z$. In this case the hypotheses reduce to: f is analytic on G, and $f' \neq 0$ on G. $D(c,r)$ and $\bar{D}(c,r)$ will

94   CHAPTER 4.   LOCAL ANALYTIC FUNCTIONS

denote the open and closed disks with center c and radius r.

LEMMA 2.5 Let $b \in G$, let $0 < m \leq |f'(b)|$, and fix ϵ, $0 < \epsilon < 1$. If δ is less than the distance from b to the complement of G and if

$$|f'(w_2)-f'(w_1)| \leq \epsilon m/\sqrt{2} \quad \text{for} \quad w_1, w_2 \in \bar{D}(b,\delta) \qquad (2.6)$$

then f is one to one on $\bar{D}(b,\delta)$ and the image of this disk contains the disk $D(a,r)$, where $a = f(b)$ and $r = (1-\epsilon)\delta m$.

Proof. (2.6) and the mean value theorem (applied separately to the real and imaginary parts of f) give

$$|f(w_2)-f(w_1)-f'(b)(w_2-w_1)| \leq \epsilon m|w_2-w_1| \quad \text{for}$$

$$w_1, w_2 \in \bar{D}(b,\delta). \qquad (2.7)$$

This implies that

$$|f(w_2)-f(w_1)| \geq (1-\epsilon)m|w_2-w_1| \quad \text{for} \quad w_2, w_1 \in \bar{D}(b,\delta) \qquad (2.8)$$

and therefore that f is one to one on $\bar{D}(b,\delta)$.

Now fix z with $|z-a| < r$, and apply the fixed point theorem to

$$F(w) = w - \frac{f(w)}{f'(a)} + \frac{z}{f'(a)} .$$

From (2.7) it follows that

SECTION 2. LOCAL SOLUTIONS TO ANALYTIC EQUATIONS

$$|F(w_2)-F(w_1)| = \left| w_2 - w_1 - \frac{f(w_2)-f(w_1)}{f'(b)} \right| \leq \epsilon |w_2 - w_1|,$$

so F satisfies the required Lipschitz condition. It must also be shown that F transforms $\bar{D}(b,\delta)$ into itself. If $|w-b| \leq \delta$, then

$$|F(w)-b| = \left| w - b - \frac{f(w)}{f'(b)} + \frac{f(b)}{f'(b)} + \frac{z}{f'(b)} - \frac{a}{f'(b)} \right|$$

$$< \epsilon |w-b| + \frac{r}{|f'(b)|} = \epsilon\delta + (1-\epsilon)\delta = \delta.$$

Because of the strict inequality in the last line, the fixed point w lies in the open disk $D(b,\delta)$, and clearly it satisfies $f(w) = z$.

Proof of Theorem 2.3. Let d be less than the distances from K_z and K_w to the complements of G_z and G_w, and let K_z' and K_w' be the sets of points at distance $\leq$ d from K_z and K_w. Let M be the max of $|\partial F/\partial z|$ on the compact set $K_z' \times K_w'$, and let 2m be the min of $|\partial F/\partial w|$ on the compact set $K_z' \times K_w' \cap N$ where N is the set of (a,b) with $F(a,b) = 0$. For arbitrary fixed $\epsilon < 1$, use the uniform continuity to find $\delta < d$ so that

$$|(\partial F/\partial z)(z_2,w_2)-(\partial F/\partial z)(z_1,w_1)| \leq \epsilon/\sqrt{2} \tag{2.9}$$

$$|(\partial F/\partial w)(z_2,w_2)-(\partial F/\partial w)(z_1,w_1)| \leq \epsilon m/\sqrt{2} \tag{2.10}$$

for $z_1, z_2 \in K_z'$, $w_1, w_2 \in K_w'$, with $|z_2-z_1| \leq \delta$ and $|w_2-w_1| \leq \delta$.

It follows from (2.10) that

$$|(\partial F/\partial w)(z,w)| \geq m \text{ if } |z-a| \leq \delta \text{ and } |w-b| \leq \delta \qquad (2.11)$$

for some $(a,b) \in K_z \times K_w \cap N$.

Let δ_1 be less than δ and less than $(1-\epsilon)m\delta/M\sqrt{2}$.

Let $(a,b) \in K_z \times K_w$, with $F(a,b) = 0$, and fix z, $|z-a| < \delta_1$. Then

$$|F(z,b)| = |F(z,b)-F(a,b)| \leq M\sqrt{2}|z-a| < (1-\epsilon)\delta m,$$

so 0 is within distance $r = (1-\epsilon)\delta|(\partial F/\partial w)(z,b)|$ of $F(z,b)$ and the lemma, applied to $f(w) = F(z,w)$, provides a solution to $F(z,w) = 0$ which lies in $D(b,\delta)$ and is the only solution that does.

From (2.9) and (2.10) it follows that

$$|F(z,w)-F(a,w)-(\partial F/\partial z)(a,b)(z-a)| \leq \epsilon|z-a| \text{ if } |z-a| \leq \delta$$
and $|w-b| \leq \delta,$ \hfill (2.12)

$$|F(a,w)-F(a,b)-(\partial F/\partial w)(a,b)(w-b)| \leq \epsilon m|w-b| \text{ if}$$

$|w-b| \leq \delta.$ \hfill (2.13)

From (2.13) it follows that if $|z-a| < \delta_1$, then

SECTION 2. LOCAL SOLUTIONS TO ANALYTIC EQUATIONS

$(1-\epsilon)m|w(z)-b| \leq |F(a,w(z))-F(a,b)|$

$= |F(a,w(z))-F(z,w(z))| \leq M\sqrt{2}|z-a|.$ (2.14)

and therefore also

$|F(a,w(z))-F(z,w(z))-(\partial F/\partial w)(a,b)(w(z)-b)| \leq$

$(\epsilon M\sqrt{2}/(1-\epsilon))|z-a|.$ (2.15)

(In both (2.13) and (2.14), $F(z,w(z)) = F(a,b) = 0$ is used.)

First, in (2.12) put $w = w(z)$, divide by $z-a$, and take the lim sup as $z \to a$. Since the lim sup is at most the arbitrary ϵ, it follows that

$\lim_{z \to a} (F(z,w(z))-F(a,w(z)))/(z-a) = (\partial F/\partial z)(a,b).$ (2.16)

Now do the same in (2.15). In view of (2.16) and the fact that the term on the right is arbitrarily small, the result is that

$\lim_{z \to a} (w(z)-b)/(z-a) = -(\partial F/\partial z)(a,b)/(\partial F/\partial w)(a,b),$

which proves the differentiability of w.

THEOREM 2.17 If F is analytic on $G_z \times G_w$, and w is analytic on G_z and takes values in G_w, then $F(z,w(z))$ is analytic on G_z. Consequently, if the equation $F(z,w(z)) = 0$ holds near one point, it holds everywhere.

Proof. With $w = w(z)$, write

$$F(z,w)-F(a,b) = F(z,w)-F(a,w) + F(a,w)-F(a,b),$$

divide by $z-a$, and let $z \to a$. By (2.16), the first term on the right has the limit $(\partial F/\partial z)(a,b)$. By (2.13), (2.14), and the differentiability of w, the second has the limit $(\partial F/\partial w)(a,b)w'(a)$.

Consider the algebroid equation

$$F(z,w) = w^m + \sum_{k=0}^{m-1} a_k(z)w^k = 0 \tag{2.18}$$

with the coefficients a_k analytic on G.

DEFINITION 2.19 A point $a \in G$ is a critical point if for some point b both F and $\partial F/\partial w$ are 0 at (a,b).

For algebroid equations, Theorems 2.3 reads as follows.

THEOREM 2.20 If a is not a critical point of F, then there are m distinct points b with $F(a,b) = 0$ and corresponding solutions w_{ab} to the equation $F(z,w) = 0$. If K is a compact set without critical points on which $|a_k(z)| \leq M$, then, for $a \in K$, the corresponding b satisfy $|b| \leq M+1$, so that the radii of convergence of the w_{ab} are all $\geq r > 0$.

Proof. The assumption that a is not a critical point implies that the equation $F(a,w) = 0$ does not have multiple roots, therefore that it has m distinct

SECTION 3.   ANALYTIC LINEAR DIFFERENTIAL EQUATIONS   99

roots.   The assertion that $|b| \leq M+1$ follows from Problem 6, Section 7 of Chapter 2.   The assertion about the radii of convergence then follows from Theorem 2.3.

For use with Cauchy's inequalities it is necessary to have an explicit and good evaluation of the radii of convergence, not just the existence of a positive lower bound.   It is shown in the next chapter that the radii of convergence of the w_{ab} are all at least the distance from a to the nearest critical point   or to the boundary of G.

In principle (and sometimes in practice) the critical points can be located by using the following theorem.

Theorem 2.21   (From Algebra)   There is   a   computable   polynomial D in the coefficients $a_0, \ldots, a_{m-1}$ such that z is a critical point if and only if

$$D(a_0(z), \ldots, a_{m-1}(z)) = 0.$$

D   is called the discriminant.   In the case of degree 2 it is the familiar $a_1^2 - 4a_0$, i.e. $b^2 - 4ac$.

Proof.   See Problems 4-6 of this section.

3.   ANALYTIC LINEAR DIFFERENTIAL EQUATIONS.

THEOREM 3.1   If f and the a_k are analytic   on   the disk $|z-a| < r_0$, and $y_0, \ldots, y_{d-1}$ are given, there is a unique analytic y on $|z-a| < r_0$ with

$$y^d(z) = \sum_{k=0}^{d-1} a_k(z) y^k(z) + f(z),$$

$$y^n(a) = y_n, \quad n = 0, \ldots, d-1 \tag{3.2}$$

Proof. To simplify the notation it is supposed that $a = 0$, which can be achieved by a translation. The first step is to find a formula for the coefficients of y on the assumption that there is an analytic solution on some (possibly smaller) disk with center $a = 0$. Let

$$a_k(z) = \sum_{n=0}^{\infty} a_{kn} z^n, \quad f(z) = \sum_{n=0}^{\infty} b_n z^n. \tag{3.3}$$

If

$$y(z) = \sum_{n=0}^{\infty} c_n z^n, \quad \text{then} \tag{3.4}$$

$$y^k(z) = \sum_{n=0}^{\infty} c_{n+k} \frac{(n+k)!}{n!} z^n, \text{ and}$$

$$a_k(z) y^k(z) = \sum_{n=0}^{\infty} d_{kn} z^n \tag{3.5}$$

with $\displaystyle d_{kn} = \sum_{j=0}^{n} c_{k+j} a_{k\ n-j} \frac{(k+j)!}{j!}.$ \quad (3.6)

SECTION 3. ANALYTIC LINEAR DIFFERENTIAL EQUATIONS

A useful special case (a majorant in disguise) occurs when $y(z) = a_k(z) = 1/(1-z)$, hence $a_k(z)y^k(z) = k!/(1-z)^{k+2}$. From Section 3, Chapter 1 it follows that c_n and a_{kn} are all 1 and that $d_{kn} = (k+n+1)!/(k+1)n!$. Therefore, (3.6) gives the identity

$$\sum_{j=0}^{n} \frac{(k+j)!}{j!} = \frac{(k+n+1)!}{(k+1)n!} . \qquad (3.7)$$

Comparison of the coefficients in (3.2) gives

$$c_{d+n} \frac{(d+n)!}{n!} = b_n + \sum_{k=0}^{d-1} d_{kn} , \qquad (3.8)$$

with

$$c_n = y^n(0)/n! = y_n/n! \quad \text{for } n \leq d-1. \qquad (3.9)$$

Formulas (3.6), (3.8), and (3.9) determine the coefficients: $c_0, \ldots, c_{d-1}$ are determined by (3.9); once $c_0, \ldots, c_{d-1+n}$ are determined, the d_{kn}, $k \leq d-1$, are determined by (3.6); then c_{d+n} is determined by (3.8). This proves the uniqueness.

To prove existence, determine the c_n as above, and define y by (3.4). If the series converges on a disk around 0, then Theorem 5.13, Chapter 2, shows that y satisfies (3.2) on that disk. The first problem is to show that the series for y does converge on some disk $|z| < r$.

LEMMA 3.10 Fix any $r_1 < r_0$, and let M be the max of $|f(z)|$ and the $|a_k(z)|$ on $|z| \leq r_1$. If $r \leq r_1$, $r \leq 1$, and $r < 1/M$, then the series for y converges on $|z| < r$.

Proof. Cauchy's inequalities give

$$|a_{kn}| \leq M/r^n \quad \text{and} \quad |b_n| \leq M/r^n. \qquad (3.11)$$

It will be shown by induction that
$$|c_n| \leq N/r^n \quad \text{if} \quad N \geq \frac{M}{1-Mr} \qquad (3.12)$$

and N is big enough so that (3.12) holds for $n \leq d-1$. Assume that (3.12) holds up to but not including some n. This assumption and (3.6) and (3.7) give that

$$|d_{kn}| \leq \frac{MN}{r^{k+n}} \sum_{j=0}^{n} (k+j)!/j! = \frac{MN(k+n+1)!}{(k+1)n!r^n} \qquad (3.13)$$

Therefore, by (3.8),

$$|c_{d+n}| \leq \frac{n!}{(d+n)!}\left(\frac{M}{r^n} + \frac{MN}{r^n}\frac{(d+n)!}{dn!}\sum_{k=0}^{d-1}\frac{1}{r^k}\right)$$

$$\leq \frac{M}{r^n} + \frac{MN}{r^{d+n-1}} \leq \frac{N}{r^{d+n}},$$

and the inductive step is established.

The inequality (3.2) shows that the series for y does converge on $|z| < r$, with r as in the Lemma. Let

SECTION 3. ANALYTIC LINEAR DIFFERENTIAL EQUATIONS 103

r_2 be the radius of convergence. It will be shown that $r_2 \geq r_1$, which will prove the theorem, since r_1 is any is any number $< r_0$.

Suppose that $r_2 < r_1$. Let r be the minimum of the numbers r_1-r_2, r_2, 1, and 1/2M, and fix an r_3 satisfying $r_2-r < r_3 < r_2$. If $|a| = r_3$, what has been proved can be applied on the disk D_a with center a and radius r to find an analytic y_a satisfying the differential equation on D_a and the initial conditions $y_a^n(a) = y^n(a)$ for $n \leq d-1$. By the uniqueness part of Theorem 3.1, $y_a(z) = y(z)$ on a disk with center a, and by Unique Continuation (Theorem 5.2 of Chapter 3) $y_a(z) = y(z)$ on $D_a \cap |z| < r_2$. If b is another point with $|b| = r_3$, and if D_a and D_b overlap, then $y_a(z) = y_b(z) = y(z)$ on $D_a \cap D_b \cap |z| < r_2$, so by Unique Continuaton again, $y_a(z) = y_b(z)$ on $D_a \cap D_b$. This means that y has an analytic extension (defined by $y(z) = y_a(z)$ if $z \in D_a$) to the disk $|z| < r_3+r$. By Theorem 4.5 of Chapter 3, the radius of convergence of the series for y must be at least $r_3+r > r_2$, which is a contradiction.

THEOREM 3.14 Let the a_k, f, and y in (3.2) be analytic on the connected open set G. If y satisfies the differential equation near one point in G, then it satisfies the differential equation everywhere.

Proof. Both sides are analytic, so unique continuation applies.

Up to the last paragraph, the proof of Theorem 3.1 provided a result similar to Theorem 2.20: the existence of an analytic solution and (not very good) lower bound for the radius of convergence. The last paragraph, which gives the best lower bound, is a simple example of an important procedure called analytic continuation - the subject of the next chapter.

The estimates in the proof of Picard's theorem can be used to provide an M for use in Cauchy's inequalities. If the coefficients and f are analytic and bounded on $D(a,r)$, let (as in Problem 2, Section 4, Chapter 2)

$$\|a\|^2 = \sup_{|z-a|<r} 1 + \sum_{k=0}^{d-1} |a_k(z)|^2, \quad \|f\| = \sup_{|z-a|<r} |f(z)| \quad (3.15)$$

First convert to a first order vector equation as in Section 4 of Chapter 2. With $Y_i = y^{i-1}$, the vector equation is

$$Y_i' = Y_{i+1} \text{ for } i < d, \quad Y_d' = \sum_{k=0}^{d-1} a_k Y_{k+1} + f \quad (3.16)$$

To simplify the notation, take $a = 0$, which can be achieved by translation.

Picard's theorem involves functions of a real

SECTION 3. ANALYTIC LINEAR DIFFERENTIAL EQUATIONS 105

variable, so for fixed z in $D(0,r)$, let $U(t) = Y(zt)$. The equations for U become

$$U_i' = zU_{i+1} \text{ for } i < d, \quad U_d' = z\left(\sum_{k=0}^{d-1} b_k U_{k+1} + g\right), \text{ with} \quad (3.17)$$

$b_k(t) = a_k(tz)$ and $g(t) = f(tz)$, i.e.

$U' = G(t,U)$ with $G_i(t,U) = zU_{i+1}$, $i < d$,

$$G_d(t,U) = z\left(\sum_{k=0}^{d-1} b_k U_{k+1} + g\right).$$

With $N = 0$, (4.7) of Chapter 2 gives

$|Y(z)-Y(0)| = |U(1)-U(0)| \leq (M_0/M_1)(e^M - 1)$, if

$|G(t,U(0))| \leq M_0$ and $|G(t,U)-G(t,V)| \leq M_1|U-V|$. (3.18)

With the aid of the Cauchy-Schwarz inequality it is seen that (3.18) holds with

$$M_0 = r(\|a\| |Y(0)| + \|f\|), \quad M_1 = r\|a\| \quad (3.19)$$

Therefore, the solution y to (1.10) satisfies

$|y(z)| \leq M$ on $|z-a| < r$, with

$$M \leq \left(|Y(0)| + \frac{\|f\|}{\|a\|}\right) e^{r\|a\|} \quad (3.20)$$

Because of the exponential in (3.20), the evaluation of

106   CHAPTER 4.   LOCAL ANALYTIC FUNCTIONS

M may increase dramatically with r - but the error decreases exponentially with the number of terms, so things are not always as bad as they seem.

Example 3.21   Compute the solution to

$y'' + (\sin x) y' + x^2 y = 0$, $y(0) = 0$, $y'(0) = 1$, with error $\leq 10^{-6}$ on $|x| \leq 1$ and on $|x| \leq 1/2$.

In the complex equation the coefficients are analytic on the whole plane, so the solution is analytic on the plane, and r can be anything > 1 in the first case and anything $> 1/2$ in the second. (Recall that the r for computational use in Cauchy's inequalities must be $> |z-a|$ for all z entering the computation.) Take $r = 2$ for the first, $r = 1$ for the second, center 0. Formula (3.19) and the evaluation of $|\sin z|$ in Problem 2, Section 4, Chapter 3 give

		$r = 2$	$r = 1$
max $|\sin z|$ =		3.6269	1.1752
$\|a\|$	$\leq$	5.4913	1.8388
$e^{r \|a\|}$	$\leq$	58842.	6.2890
M	$\leq$	58842.	6.2890
error in stopping at deg N	$\leq$	$58842/2^N$	$6.2890/2^N$
value of N for error $\leq 10^{-6}$	=	36	23

CHAPTER 4. LOCAL ANALYTIC FUNCTIONS - PROBLEMS

PROBLEMS

SECTION 1.

1. Evaluate $L_a(1+i)$, $a = 1, i, -1, -i$
2. What are the ranges of L_a, $a = 1, i, -1, -i$?
3. What is the range of L_1 on $|z-a| < r$, z not a non-positive real,?

SECTION 2.

1. Find the critical points of $w^2 = z$ on $G = R^n$
2. Show that the equation $w^2 = z$, $w(a) = a^2$, has two (everywhere) distinct solutions on the disk with center a and radius = the distance from a to the nearest critical point.
3. Suppose the algebroid equation $F(z,w) = 0$ has no critical point in the connected open G. Let u and v be two solutions to $F(z,w(z)) = 0$ on G. Show that $u = v$ on G if $u(z_0) = v(z_0)$ for any one point z_0.
4. (Euclidean Algorithm) Let p_0 and p_1 be polynomials. There are unique polynomials q and r such that

$p_0 = p_1 q + r$, with degree r < degree p_1 (or $r = 0$).

Hint. Let the leading terms in p_0 and p_1 be $a_n z^n$ and $b_m z^m$. If $m < n$, set $r_1 = p_0 - (a_n/b_m) z^{n-m} p_1$. If deg r_1 < deg p_1, you are done. If not, repeat, with p_0 replaced by r_1.

5. (The discriminant) Let p have leading coefficient 1, with the other coefficients $a_0, \ldots, a_{d-1}$ undetermined. Setting $p_0 = p$, $p_1 = p'$, form a sequence p_k so that $p_{k-1} = p_k q_k + p_{k+1}$ by the Euclidean Algorithm. The p_k are polynomials in w with coefficients that are rational functions of $a_0, \ldots, a_{d-1}$. Write the first one that is independent of w as D/E, where D and E are polynomials with no common factor. Then D is the discriminant. Show that for a particular choice of the coefficients $a_0, \ldots, a_{d-1}$, p and p' have a common zero if and only if $D = 0$.

6. While the calculation of discriminants is constructive, it is a mess. Calculate D for p of degree 3. On the 4th try I got

$$D = 2a_1 D_2 - D_1(4a_2 D_2 - 3D_1) \text{ with } D_1 = 9a_0 - a_1 a_2, \; D_2 = 3a_1 - a_2^2$$

(but maybe a 5th try would have been advisable). Since it is a polynomial of degree 5 in the coefficients, which are analytic functions themselves, the discriminant can be used to locate the critical points only in very special situations.

7. If the coefficients are analytic on G, there are only a finite number of critical points in any compact subset of G.

8. If f is analytic on a disk with center a then, for some positive integer m, $f(z) - f(a) = g(z)^m$, where g is analytic and $g'(a) \neq 0$.

9. If f is analytic on the open set G then f(G) is open. (Use Exercise 8.)

CHAPTER 5. ANALYTIC CONTINUATION

1. ANALYTIC CONTINUATION ALONG PATHS.

The last two sections contained proofs of the local existence of analytic functions satisfying either an analytic equation or an analytic linear differential equation. For each point a in a connected open set G they provided a number of analytic functions g_a satisfying the equation on some disk with center a. The problem of analytic continuation is to decide whether, for each a, it is possible to choose one of the g_a so that they fit together to determine an analytic function on all of G. If, for example, G is the complement of 0, and the L_a are the branches of the logarithm, it is not possible to piece them together to form an analytic function on G: if it were, 1/z would have a primitive on G, so its integral over a circle around 0 would have to be 0. More generally, the problem of analytic continuation is as follows: given a connected open set G, a point a ∈ G, and a function f, analytic on a neighborhood of a, is there an F, analytic on G and coinciding with f on a neighborhood of a? It is approached via the study of analytic continuation along paths. To avoid subscripts, the paths are usually defined on [0,1], and G always stands for an open connected set.

SECTION 1. ANALYTIC CONTINUATION ALONG PATHS

DEFINITION 1.1 An analytic continuation along a path p is a family of power series F_t such that

1) F_t is a power series with center $p(t)$, radius $r(t)$, and disk of convergence $D(t)$.

2) There is a number $\delta_0 > 0$ such that if $|t-s| \leq \delta_0$, then $F_t = F_s$ on $D(t) \cap D(s)$.

If f is analytic on a neighborhood of some $p(s)$, and F_s is the power series of f with center $p(s)$, then $\{F_t\}$ is the analytic continuation of f along p.

The connection between local and global analytic functions on the one hand, and analytic continuation along paths on the other, is as follows.

THEOREM 1.2 Let f be analytic on a neighborhood of a point $a \in G$. There is an analytic F on all of G with F = f on a neighborhood of a if and only if

1) For each path p in G with initial point a, f has an analytic continuation along p.

2) If $\{F_t\}$ and $\{G_t\}$ are analytic continuations of f along paths p and q with $p(1) = q(1)$, then $F_1 = G_1$.

Proof. If F exists, let F_t be the power series for F at $p(t)$. If 1) and 2) hold, let z be any point in G, let p be a path from a to z, and let $\{F_t\}$ be an analytic continuation of f along p. Define $F(z) = F_1(z)$, which, according to 2) is independent of the particular path and continuation chosen. With this definition it follows that $F(w) = F_1(w)$ on any disk with center z and contained in $G \cap D(1)$, so F is analytic.

The theorem separates the problem of continuing a given f into two parts of different natures: to show that continuation along all paths is possible; and to show that the outcome depends only on the endpoints, not on the particular path joining them. The role of paths is that when analytic continuation along a given path is possible, it is unique.

THEOREM 1.3 If $\{F_t\}$ and $\{G_t\}$ are analytic continuations of f along the same path p, then $F_t = G_t$ for all t.

Proof. Let $F_0 = G_0 =$ the series of f at $p(0)$ and let s = the sup of those t_1 such that $F_t = G_t$ for $0 \leq t \leq t_1$. Choose $\delta > 0$ so that if $|t-s| \leq \delta$, then $p(t) \in D_s$ and so that $F_t = F_s$ on $D_t \cap D_s$ and $G_t = G_s$ on $D_t \cap D_s$, where D_t is the smaller of the disks for F_t and G_t. If $t < s$, then $F_s = F_t = G_t = G_s$ on $D_t \cap D_s$, so by unique continuation, $F_s = G_s$. Reversal of the argument shows that if $t > s, |t-s| \leq \delta$, then $F_t = G_t$. Consequently, the sup s must be 1.

SECTION 1. ANALYTIC CONTINUATION ALONG PATHS

LEMMA 1.4 If $\{F_t\}$ is an analytic continuation, then $|r(t)-r(s)| \leq |p(t)-p(s)|$ whenever $|t-s| \leq \delta_0$. Therefore, $r(t)$ is continuous, and there exists $r_0 > 0$ such that

$$r(t) \geq r_0 \text{ for all } t. \tag{1.5}$$

Proof. Suppose that $r(t) \geq r(s)$. If $p(s) \notin D(t)$, then $|p(t)-p(s)| \geq r(t) \geq |r(t)-r(s)|$. If $p(s) \in D(t)$, then F_s is the power series for F_t centered at $p(s)$, and this series converges on $D(t)$.

THEOREM 1.6 Let f be analytic on a neighborhood of a point a, and let p be a path with initial point a. f has an analytic continuation along p if and only if there is a positive number r such that in every "partial continuation" on a sub-interval $0 \leq t \leq s$, the radii $r(t)$ are all $\geq r$.

Proof. The necessity is shown by the lemma. To prove the sufficiency choose δ so that if $|t-s| \leq \delta$, then $|p(t)-p(s)| < r$. Let $\{F_t\}$ be a continuation along p defined for $0 \leq t \leq s$. For $t > s$, $t-s \leq \delta$, let F_t be the power series for F_s centered at $p(t)$. This extends the definition of $\{F_t\}$ by the fixed amount δ.

As a first application, consider the differential equation

$$y^d = \sum_{k=0}^{d-1} a_k(z) y^k + f \ . \qquad (1.7)$$

THEOREM 1.8 Let the a_k and f be analytic on G, let y be an analytic solution in a neighborhood of some point a, and let p be a path in G with initial point a. Then y has an analytic continuation $\{y_t\}$ along p, and all y_t are solutions to the differential equation.

Proof. Let $\{y_t\}$ be a partial continuaton defined for $t \leq s$, so that these y_t satisfy the differential equation. By Theorem 3.1, Chapter 4, $r(t) \geq r =$ the distance from the range of p to the complement of G. The extension of $\{y_t\}$ by $y_t =$ the power series of y_s centered at $p(t)$ provided by Theorem 1.6 obviously satisfies the differential equation and extends $\{y_t\}$ by the fixed amount δ.

As a second example, consider the algebroid equation

$$F(z,w) = w^m + \sum_{k=0}^{m-1} a_k(z) w^k = 0 \qquad (1.9)$$

THEOREM 1.10 Let the a_k be analytic on G, and let there be no critical point in G. If w is an analytic solution to (1.9) on a neighborhood of some point a, and p is any path in G with initial point a, then w has

SECTION 1.   ANALYTIC CONTINUATION ALONG PATHS   115

an analytic continuation $\{w_t\}$ along p, and all the w_t satisfy (1.9).

Proof. Let K be the compact range of p. By Theorem 2.20 of Chapter 4 the radii of convergence of the local solutions have a positive lower bound for a $\in$ K. If $\{w_t\}$ is a partial continuation defined on t $\leq$ s with each w_t satisfying (1.9), the extension of $\{w_t\}$ provided by Theorem 1.6 clearly continues to satisfy (1.9).

Some results for the general analytic equation F(z,w) = 0 will be presented via exercises and comments. The situation is similar, but not identical, to that in Theorem 1.10. In particular, the lower bound for the radii in a partial continuation is no longer automatic. One way to obtain it is to restrict to functions F with the following property.

DEFINITION 1.11   F is proper if for each compact $K_z \subset G_z$, the set of w satisfying F(z,w) = 0 for some $z \in K_z$ is compact.

THEOREM 1.12   If F is proper, Theorem 1.11 holds.

Proof. See Problem 4.

The assumption that f is proper is often too strong. For example, F(z,w) = e^w-z is not proper: for each z $\neq$ 0, the corresponding set of w is unbounded. In some cases the required lower bound on the radii can be obtained in other ways. For example

THEOREM 1.13 Let f be analytic and $\neq 0$ on G. Let w be analytic on a neighborhood of a and satisfy $e^w = f$, and let p be any path with initial point a. Then w has an analytic continuation along p and each w_t satisfies the equation.

Proof. See Problem 5.

2. THE MONODROMY THEOREM.

Let f be analytic on a neighborhood of the point $a \in G$ (open and connected, as always). Theorem 1.2 provides two conditions for the existence of an F which is analytic on G and coincides with f on a neighborhood of a: it must be possible to continue f along any path in G with initial point a; and the result of such continuation must depend only on the end point of the path. For algebroid functions and for solutions to linear analytic differential equations, the first condition was established in the last section. The second condition is the subject of the Monodromy Theorem. For simplicity of notation, all paths are defined on $[0,1]$. The distance between paths p and q is the number

$$d(p,q) = \sup\{|p(t)-q(t)| : 0 \leq t \leq 1\} \qquad (2.1)$$

THEOREM 2.2 Let f be analytic on a neighborhood of a and let $\{F_t\}$ be an analytic continuation of f

SECTION 2. THE MONODROMY THEOREM

along a path p with initial point a and with $r(t) \geq r_0$ for all t. If $d(q,p) < r_0/2$, then f has an analytic continuation along q defined by G_t = the series of F_t centered at $q(t)$.

Proof. Choose δ so that if $|s-t| \leq \delta$, then $|p(s)-p(t)| < r_0/2$, $|q(s)-q(t)| < r_0/2$, and $F_t = F_s$ on $D(t) \cap D(s)$.

Suppose that $\{G_t\}$, as defined, is a continuation on $t \leq s$, and let $|t-s| < \delta$. Then $q(t)$ lies in $D(t) \cap D(s)$, so $G_t = F_t = F_s$ on this intersection, therefore on the disk with center $q(t)$ and radius $r_0/2$. Since G_s is also equal to F_s on this disk, it follows that G_t is the series for G_s, so that $\{G_t\}$ remains a continuation on $t < s+\delta$, with δ independent of s.

DEFINITION 2.3 Let p and q be paths in G with the same endpoints a and b. A homotopy from p to q in G is a family $\{p^s\}$, $0 \leq s \leq 1$, of paths in G such that

a) $p^0 = p$ and $p^1 = q$.
b) $p^s(0) = a$ and $p^s(1) = b$ for each s.
c) The function $h(s,t) = p^s(t)$ is continuous on $[0,1] \times [0,1]$.

G is simply connected if it is connected and every closed path is homotopic to a constant path.

THEOREM 2.4 If G is simply connected, then any

two paths in G with the same endpoints are homotopic.

Proof. Given p and q, define the p^s as follows.
For $0 \leq s \leq 1/4$,
$p^s(t) = p(t/(1-s))$ if $t \leq 1-s$, $p^s(t) = b$ if $t > 1-s$.
For $3/4 \leq s \leq 1$,
$p^s(t) = q(t/s)$ if $t \leq s$, $p^s(t) = b$ if $t > s$.
For $1/4 \leq s \leq 1/2$,
$p^s(t) = p^{1/4}(t(2-4s))$ if $t \leq 3/4$.
For $1/2 \leq s \leq 3/4$,
$p^s(t) = p^{3/4}(t(4s-2))$ if $t \leq 3/4$.

This leaves $p^s(t)$ open for $1/4 < s < 3/4$ and $t > 3/4$. The path $P(s) = p^s(3/4)$, $1/4 \leq s \leq 3/4$ is closed, taking the value b at both ends. Since G is simply connected, there is a function $H(s,t)$ on $1/4 \leq s \leq 3/4$, $3/4 \leq t \leq 1$, such that $H(s,3/4) = P(s)$, $H(s,1) = H(1/4,t) = H(3/4,t) = b$. Set $p^s(t) = H(s,t)$. (Note the reversal of the roles of s and t here.)

THEOREM 2.5 (Monodromy Theorem) Let $\{p^s\}$ be a homotopy between paths p and q in G with endpoints a and b, and let f be analytic on a neighborhood of a. If, for each s, f has an analytic continuation $\{F_t^s\}$ along p^s, then $F_1^0 = F_1^1$.

THEOREM 2.6 (Monodromy Theorem) Let G simply connected, and let f, analytic on a neighborhood of a, have an analytic continuation along every path with

SECTION 2. THE MONODROMY THEOREM

initial point a. Then there is an F which is analytic on G and coincides with f on a neighborhood of a.

Proof. Theorem 2.6 follows immediately from Theorems 2.4, 2.5 and 1.2, so it is a question of proving Theorem 2.5.

Fix u, and let $r_o \leq r^u(t)$ for all t. Using the uniform continuity of $h(s,t) = p^s(t)$, find δ so that $d(p^s, p^u) < r_o/2$ if $|s-u| \leq \delta$. Theorem 2.2 shows that if $|s-u| \leq \delta$, then F_1^s is the series for F_1^u with center at $p^s(1) = p^u(1) = b$. Since any small change in s does not change F_1^s, neither does a large change, and F_1^s remains the same throughout the homotopy.

In the light of the Monodromy Theorem the results on the solutions of algebroid equations and analytic linear differential equations appear as follows.

THEOREM 2.7 In the differential equation (1.7), let the a_k and f be analytic on the simply connected G. If $y_o, \ldots, y_{d-1}$ and a are given, there is one and only one function y which is analytic on G and satisfies (1.7) and $y^k(a) = y_k$, $0 \leq k \leq d-1$.

THEOREM 2.8 In the algebroid equation (1.9), let the a_k be analytic on a simply connected G without critical points. Then there are m distinct functions y_j that are analytic on G and satisfy (1.9). If $j \neq k$, then $y_j(z) \neq y_k(z)$ for every $z \in G$.

THEOREM 2.9 If f is analytic and $\neq 0$ on the simply connected G, then there is an analytic log f on G, i.e., L which is analytic on G and satisfies $e^{L(z)} = f(z)$. Any two such functions differ by an integer multiple of $2\pi i$. The functions

$$w_j(z) = e^{(L(z)+j2\pi i)/m}, \quad j = 0, \ldots, m-1$$

are the distinct functions such that $w^m = f$, i.e. are the analytic m-th roots of f.

Proof. Theorems 1.13 and 2.6

Example 2.10 Calculate, with error $\leq 10^{-6}$,

$$\int_0^1 f(x)dx \quad \text{for} \quad f(x) = x\sqrt{1+e^x}.$$

The complex function f satisfies $w^2 - z^2(e^z+1) = 0$. This is no good, however, since 0 is a critical point and also one of the integration limits. However, $f(z) = zg(z)$, where g satisfies the equation $F(z,w) = w^2 - (e^z+1) = 0$. In this case the critical points are the odd multiples of πi, so by Theorem 2.8, g is analytic on the disk $|z-1/2| < \sqrt{\pi^2+1/4}$. (The theorem gives two such analytic functions; g is the one with the value $\sqrt{2}$ at $z = 0$, the other has the value $-\sqrt{2}$ at $z = 0$.) On the disk $|z-1/2| < \pi$,

$$|e^z+1| \leq e^{|z|}+1 \leq e^{\pi+1/2}+1 = 39.17, \text{ so}$$

SECTION 2. THE MONODROMY THEOREM

$|g(z)| \leq \sqrt{39.17} = 6.26$.

If $g(z) = \sum_{n=0}^{\infty} a_n(z-1/2)^n$, then $|a_n| \leq 6.26/\pi^n$.

Therefore,

$$f(z) = zg(z) = ((z-1/2)+1/2))g(z) = \sum_{n=0}^{\infty} b_n(z-1/2)^n \text{ with}$$

$b_o = a_o/2$ and $b_n = a_{n-1}+a_n/2$, so $|b_n| \leq 22.8/\pi^n$. (2.11)

Since

$$\int_{1/2-y}^{1/2+y} f(x)dx = 2 \sum_{n=0}^{\infty} b_{2n}/(2n+1) \, y^{2n+1},$$

with $y = 1/2$, the error in stopping at $n = N$ is at most

$$22.8/(2N+3) \sum_{n=N+1}^{\infty} (1/2\pi)^{2n} \leq 10^{-6} \text{ for } N = 3.$$

The coefficients a_n, $n \leq 6$, are determined by the equation

$$(T_{1/2}^6 g)^2 = T_{1/2}^6 (e^z+1) \mod (x-1/2)^7,$$

then the b_n by (2.11).

3. CAUCHY'S INTEGRAL FORMULA AND THEOREM.

Theorem 1.8 and the Monodromy Theorem provide the general form of Cauchy's Integral Theorem and Formula. First, Theorem 1.8 allows a different interpretation of path integrals.

THEOREM 3.1 Let f be analytic on G and let p be a path with end points $a = p(0)$ and $b = p(1)$. Let $\{F_t\}$ be an analytic continuation of a primitive of f along p. Then

$$\int_p f(z)dz = F_1(b) - F_0(a) .$$

Proof. By Theorems 3.4 and 3.5 of Chapter 3, f has a primitive F on a neighborhood of a. This means that $F' = f$ on a neighborhood of a, and Theorem 1.8 guarantees the existence of the analytic continuation, with each F_t continuing to satisfy this equation. It follows that the function $G(t) = F_t(p(t))$ is a primitive of $f(p(t))p'(t)$, hence that the integral is $G(1) - G(0)$.

The general form of Cauchy's Integral theorem is:

THEOREM 3.2 Let f be analytic on the simply connected G. If p is any closed path in G, then the integral of f over p is 0.

SECTION 3.   CAUCHY'S INTEGRAL FORMULA AND THEOREM

Proof. Let $a = p(0) = p(1)$, and let F be a primitive of f on a neighborhood of a. Let F_t and G_t be the analytic continuations of F along p and along the path $q(t) = a$. By definition $F_0 = G_0 = G_1 = F$, and by the Monodromy Theorem, $F_1 = G_1$. Hence $F_1(a)-F_0(a) = G_1(a)-G_0(a) = 0$.

The general form of Cauchy's Integral Formula is:

THEOREM 3.3 Let f be analytic on the simply connected G. If a is a point in G and p is a closed path in G that does not pass through a, then

$$f(a) = (1/2\pi i)\left(\int_p f(z)/(z-a)\ dz\right)\Big/\mathrm{ind}_a p \qquad (3.4)$$

$$\mathrm{ind}_a p = (1/2\pi i)\int_p 1/(z-a)\ dz \qquad (3.5)$$

Proof. Apply the Integral Theorem to the analytic function $g(z) = (f(z)-f(a))/(z-a)$.

PROBLEMS

SECTION 1.

1. Let $\{F_t\}$ be the analytic continuation of L_1 (the logarithm) along $p(t) = e^{it}$, $0 \leq t \leq 2\pi$. What is $F_{2\pi}$?

2. Let w be the analytic solution to $w^2 = z$, $w(1) = 1$, on a disk around 1, and let $\{F_t\}$ be the analytic continuation along $p(t) = e^{it}$, $0 \leq t \leq 2\pi$. What is $F_{2\pi}$? Let $\{F_t\}$ be the analytic continuation along $p(t) = e^{it}$, $0 \leq t \leq 2n\pi$. What is $F_{2n\pi}$?

3. Let w be the analytic solution to $w^2 = z(2-z)$, $w(1) = 1$, and let $\{E_t\}$, $\{F_t\}$, $\{G_t\}$, $\{H_t\}$ be the analytic continuations along $p(t) = e^{it}$, $q(t) = 2 + e^{it}$, qp, and $q^{-1}p$, $0 \leq t \leq 2\pi$. What are $E_{2\pi}$, $F_{2\pi}$, $G_{4\pi}$, and $H_{4\pi}$?

4. Prove Theorem 1.12.

5. Prove Theorem 1.13 Hint: $w' = f'/f$.

SECTION 2.

All homotopies take place in a fixed connected open set G. The statement "p is homotopic to q in G" is abbreviated by $p \sim q$.

1. If $p_1 \sim p_2$ and $p_2 \sim p_3$, then $p_1 \sim p_3$.

2. Suppose that for a given closed path p there is a family $\{p_s\}$ such that

a) $p_0 = p$ and p_1 is constant.
b) Each p_s is closed.
c) The function $h(s,t) = p_s(t)$ is continuous on $[0,1] \times [0,1]$.

Then p is homotopic to a constant path. Therefore, in the definition of "simply connected" it is not necessary that all paths in the homotopy have the same endpoints, but only that all paths be closed.

DEFINITION. G is star shaped with respect to the point a in G if for each z in G the segment [a,z] lies entirely in G; G is star shaped if it is star shaped with respect to one of its points.

3. G is convex if it is star shaped with respect to all of its points.

4. If G is star shaped, it is simply connected.

5. If G is the plane with a ray removed, then G is star shaped.

6. If G is the plane with two non-intersecting rays removed, then G is simply connected. (If the rays are not parallel, G is star shaped.) What if G is the plane with several non-intersecting rays removed? Hint Problem 6 of the next section.

7. Find the critical points and the radius of convergence of the series centered at 0 when 0 is not a critical point.

a) $w^2 - \cos z = 0$, $w(0) = 1$.

b) $w^2+(\sqrt{\cos z})w+1 = 0$, $\sqrt{\cos z}$ determined by a).

c) $w^4-(\sin z+\cos z)w^2+\sin z\cos z = 0$.

d) $w^2+(\sqrt{z^2+1})w-z = 0$, $\sqrt{z^2+1} = -1$ at $z = 0$.

e) $w^2+(1/\cos z)w-1 = 0$.

8. In each of the above cases, discuss the analytic continuation of the solution w along a circle that surrounds but does not pass through a critical point, and along a circle that surrounds two critical points.

9. In each of the above cases find a (large) simply connected region in which the solution to the equation exists and is analytic.

10. In each case (except c)) let 2r be the distance from 0 to the nearest critical point. Find N giving an error $< 10^{-6}$ for the integral of w from 0 to r.

SECTION 3.

Let p be a closed path with $p(0) = p(1) = b$, and let a be a point not on p. Let f be a branch of $\log(z-a)$ on a neighborhood of b, and let $\{L_t\}$ be an analytic continuation of f along p.

1. $\text{ind}_a p = (1/2\pi i)(L_1(b)-L_0(b))$.

2. $\text{ind}_a p$ is an integer.

3. If G is any connected set containing no point of p, then $\text{ind}_a p$ is independent of the point a in G.

4. If p and q are homotopic in the complement of a, then $\text{ind}_a p = \text{ind}_a q$.

(From an intuitive point of view, $\text{ind}_a p$ is the number of times p winds counter-clockwise around a minus the number of times it winds around clockwise.)

THEOREM. The following are equivalent.

a) G is simply connected.

b) The integral of any analytic function on G along any closed path in G is 0.

c) Every analytic function on G has a primitive.

d) For every analytic $f \neq 0$ on G there is a branch of $\log f(z)$ on G.

e) For every $a \notin G$ and every path p in G, $\text{ind}_a p = 0$.

5. Prove that each of the above statements implies the next. Some of the reverse implications are not so easy. Identify the easy ones and prove them.

6. As a corollary to "e) implies a)", prove the following: If G_1 and G_2 are simply connected and $G_1 \cap G_2$ is connected, then $G_1 \cap G_2$ is simply connected.

INDEX

Absolutely convergent series, 36

Algebroid, 18, 92

Algebroid equations on simply conn. sets, 119

Alternating series, 41

Analytic cont. of primitives and path integrals, 122

Analytic cont. of sols. to alg. eq., 114

Analytic cont. of sols. to diff. eq., 114

Analytic continuation, 110

Analytic continuation (existence), 113

Analytic continuation (uniqueness), 112

Analytic continuation along a path, 111

Analytic differential equations (local solutions), 99

Analytic equations (local solutions), 92

Analytic function, 56

Analytic identities, 58

Analyticity of power series, 53, 56

Argument, 87

Branch of the logarithm, 88

Binomial coefficient, 11

Boundary, 69

Cauchy sequence, 30, 33

Cauchy's inequalities, 61, 80

Cauchy's integral formula, 76, 78, 123

Cauchy's integral theorem, 72, 122

INDEX

Cauchy-Schwarz inequality, 63

Combinations of analytic functions, 60

Combinations of Taylor polynomials, 14

Comparison test, 39

Complex Derivative, 20

Complex differentiable, 20

Complex Taylor polynomial, 19

Connected open set, 56

Convergent sequence, 30, 33

Convergent series, 35

Convex, 72

Cosine, 4

Critical point, 98

Diameter of a set, 75

Disk of convergence, 49

Differentiability of power series, 50, 52

Differential equation, 18, 42

Differential equations on simply conn. sets, 119

Discriminant, 99

Distance between paths, 116

Euclidean algorithm, 107

Exponential, 4

Fixed point theorem, 92

Geometric sum, 6

Homotopy, 117

Index of a path with respect to a point, 123

Inf, 31

Integral test, 37

Leibnitz's formula, 16, 17

Lim inf, 31

Lim sup, 31

Limit under the integral, 46

Lipschitz, 48, 92

Logarithm, 87, 120

Maximum modulus theorem, 80

Mean value theorem, 1

Minkowski's inequality, 63

Mod, 14

Monodromy theorem, 118

nth root test, 40

nth roots, 120

Open set, 56

Partial fractions, 13

Path (arc of circle), 69

Path (boundary of triangle), 69

Path (inverse), 68

Path (product), 68

Path (segment), 68

Path (smooth), 70

Path integral, 71, 122

Paths, 68

Permanence of analytic differential relations, 103

Permanence of analytic relations, 97

Picard's theorem, 43

Power series, 48

Proper, 115

Radius of convergence, 49

Ratio test, 40

Rational function, 13

Remainder, 2

Series expansion of differentiable function, 78

Simply connected, 117

Sine, 4

Star-shaped, ??

Sup, 31

Taylor polynomial, 1

Taylor's formula, 3

Uniform convergence, 45

Unique continuation, 57

Vector differential equations, 42, 48, 104